Flutter Introduction and Practice

Flutter 技术入门与实战

亢少军 编著

机械工业出版社
China Machine Press

图书在版编目（CIP）数据

Flutter 技术入门与实战 / 亢少军编著 . —北京：机械工业出版社，2019.1（2019.5 重印）
（实战）

ISBN 978-7-111-61797-6

I.F⋯ II. 亢⋯   III. 移动终端 – 应用程序 – 程序设计   IV. TN929.53

中国版本图书馆 CIP 数据核字（2018）第 303179 号

Flutter 技术入门与实战

出版发行：机械工业出版社（北京市西城区百万庄大街 22 号 邮政编码：100037）
责任编辑：吴 怡   责任校对：殷 虹
印　　刷：北京市兆成印刷有限责任公司   版　次：2019 年 5 月第 1 版第 3 次印刷
开　　本：186mm×240mm 1/16   印　张：21.75
书　　号：ISBN 978-7-111-61797-6   定　价：79.00 元

凡购本书，如有缺页、倒页、脱页，由本社发行部调换
客服热线：（010）88379426 88361066   投稿热线：（010）88379604
购书热线：（010）68326294 88379649 68995259   读者信箱：hzit@hzbook.com

版权所有·侵权必究
封底无防伪标均为盗版
本书法律顾问：北京大成律师事务所 韩光 / 邹晓东

前言

现在，主流的移动开发平台是 Android 和 iOS，每个平台上的开发技术不太一样，针对每个平台开发应用需要特定的人员，但这样一来开发效率低下，因而需要进行跨平台开发。跨平台技术从最开始的 Hybrid 混合开发技术，到 React Native 的桥接技术，一直在演进。

Hybrid 开发主要依赖于 WebView。但 WebView 是一个重量级的控件，很容易产生内存问题，而且复杂的 UI 在 WebView 上显示的性能不好。React Native 技术抛开了 WebView，利用 JavaScript Core 来做桥接，将 JavaScript 调用转为 Native 调用。React Native 最终会生成对应的自定义原生控件。这种策略将框架本身和 App 开发者绑在系统的控件上，不仅框架本身需要处理大量平台相关的逻辑，随着系统版本变化和 API 的变化，开发者可能也需要处理不同平台的差异，甚至有些特性只能在部分平台上实现，这样使得跨平台特性大打折扣。

Flutter 是最新的跨平台开发技术，可以横跨 Android、iOS、MacOS、Windows、Linux 等多个系统。Flutter 采用了更为彻底的跨平台方案，即自己实现了一套 UI 框架，然后直接在 GPU 上渲染 UI 页面。

笔者最早接触的跨平台技术是 Adobe Air 技术，写一套 Action Script 代码可以运行在 PC、Android 及 iOS 三大平台上。目前，笔者与朋友开发视频会议产品，需要最大化地减少前端的开发及维护工作量，所以，我们先后考察过 Cordova、React Native 及 Flutter 等技术。我们觉得 Flutter 方案更加先进，效率更高，后来就尝试用 Flutter 开发了全球第一个开源的 WebRTC 插件（可在 GitHub 上搜索 Flutter WebRTC）。

写作本书的目的是想传播 Flutter 知识（因为 Flutter 确实优秀），想为 Flutter 社区做点贡献的同时也为我们的产品打下坚实的技术基础。通过写作本书，笔者查阅了大量的资料，使得知识体系扩大了不少，收获良多。

本书主要内容

第 1 章介绍 Flutter 的基本概念，并搭建第一个 Flutter 程序，来感受一下 Flutter 之美。

第 2 章介绍几个重要知识点，如入口程序、Material Design、Flutter 主题、无状态组件和有状态组件、使用包资源、Http 请求。

第 3 章简单介绍 Dart 语言。Dart 语言是 Flutter SDK 指定的语言，我们很有必要补充一下它的基础知识，包括语法特性、基本语句、面向对象等。

第 4 章介绍常用组件。Flutter 里有一个非常重要的核心理念：一切皆为组件，本章主要讲解开发中用得最频繁的组件，如容器组件、图片组件、文本组件、图标组件和表单组件等。

第 5 章介绍 Material Design 风格的组件，Material Design 风格是一种非常有质感的设计风格，并提供一些默认的交互动画。本章将分类介绍这些组件。

第 6 章介绍 Cupertino 风格的组件，这是一类 iOS 风格的组件，如 CupertinoTabBar、CupertinoPageScaffold、CupertinoTabScaffold、CupertinoTabView 等。

第 7 章介绍页面布局的基础知识和技巧，如基础布局处理、宽高尺寸处理、列表及表格布局等，最后通过一个综合布局示例来演示如何编写复杂的页面。

第 8 章介绍如何处理手势，如轻击、拖动和缩放等。Flutter 中提供 GestureDetector 进行手势检测，并为手势检测提供了相应的监听。

第 9 章介绍如何加载、处理、展示资源和图片，如添加资源和图片、自定义字体等。

第 10 章介绍路由及导航是如何处理的，包括页面的渲染以及数据传递。

第 11 章介绍组件装饰和视觉效果的处理，如 Opacity（透明度处理）、DecoratedBox（装饰盒子）、RotatedBox（旋转盒子）、Clip（剪裁处理）和 CustomPainter（自定义画板）。

第 12 章介绍动画效果的制作，包含两个动画组件的使用：用 AnimatedOpacity 实现渐变效果、用 Hero 实现页面切换动画。

第 13 章介绍 Flutter 插件开发的入门知识。Flutter 插件可以和原生程序打交道，比如调用蓝牙、启用 WIFI、打开手电筒，等等。

第 14 章介绍开发工具及使用技巧，介绍几款常用的 IDE 工具，从代码的编写、辅助功能、程序调试、性能分析等多方面讲解工具及使用技巧。

第 15 章介绍测试与发布应用，包括：测试应用、发布 Android 版和 iOS 版 App。

第 16 章通过一个综合案例介绍如何使用 Flutter 实现即时通讯 App 的界面。

阅读建议

本书是一本基础入门加实战的书籍，既有基础知识，又有丰富示例，包括详细的操作步骤，实操性强。由于 Flutter 大量使用组件，所以本书对组件的讲解很详细，包括基本概念、属性及代码示例。每个组件都配有小例子，力求精简，还提供了配套网站提供完整代码，复制完整代码就可以立即看到效果。这样会给读者信心，在轻松掌握基础知识的同时快速进入实战。

如果你正在使用类似 React Native 等跨平台的技术，那么学习 Flutter 相对轻松很多。使用 Flutter 构建应用时，需要 Android 和 iOS 知识，因为 Flutter 依赖移动操作系统提供众多功能和配置。Flutter 是一种为移动设备构建用户界面的新方式，但它有一个插件机制可与 Android 和 iOS 进行数据及任务通信。本书有一章专门讲解 Flutter 的插件开发技术，可以作为进一步学习的起点。

Flutter 引用了大量 Web 开发的知识，比如 FlexBox 布局方式、盒模型等，这些都引用了 CSS 的一些思想。所以从 UI 界面的实现角度来说，只需要熟悉 Dart 语言就能轻松上手 Flutter。本书专门有一章介绍 Dart 语言的基础知识。

建议读者在一开始先把第 1～3 章基础理论通读一遍。到第 4 章时实际操作并运行每个例子，这样就能开发 Flutter 最简单的界面了。

第 5 章和第 6 章通读即可。第 7 章介绍的构建页面布局都是开发中常用的布局方法，建议读者仔细阅读、实际操作并运行每个例子。尤其是最后的布局综合例子，按步骤都走一遍，就能理解布局的思路。

第 8～12 章涵盖 Flutter 的高级用法，在开发中也经常使用。可以根据实际项目开发和学习的需要阅读。第 13 章介绍 Flutter 插件开发。这需要具备原生开发的知识，比如 Java、Objective-C 等相关知识。如果读者不需要开发插件可以略过。第 14～16 章实操性很强，读者只要根据书中的步骤仔细操作，就能快速掌握。

关于随书代码

本书所列代码力求完整，但由于篇幅所限，代码没有全放在书里。完整代码在以下网址：

http://www.flutter100.net

https://github.com/kangshaojun

致谢

首先感谢机械工业出版社吴怡编辑的耐心指点，以及推动了本书的出版。

感谢朋友段伟伟工程师，江湖人称"鱼老大"，国内骨灰级 WebRTC 开发者、视频会议产品合作者。在这里感谢鱼老大的技术分享及帮助。

感谢陈波及陈志红两位好友。在本书交稿压力最大的时侯，从内容安排及语言润色方面，都提供了一些非常有用的建议。还感谢高文翠老师对本书大纲的指导。

最后还要感谢我的家人。感谢我的母亲及妻子，在我写作过程中承担了全部的家务并照顾小孩儿，使我可以全身心地投入写作工作。在写作那段时间，正好家里阁楼装修，感谢我的父亲，他亲自管理装修工程，帮我节省了很多时间和精力。我爱你们，和你们在一起是幸运的事情！

亢少军

2018 年 12 月 7 日

目 录

前言

第1章 开启Flutter之旅 ·················· 1

1.1 Flutter 的特点与核心概念 ············ 1
    1.1.1 一切皆为组件 ·················· 2
    1.1.2 组件嵌套 ························ 2
    1.1.3 构建 Widget ··················· 3
    1.1.4 处理用户交互 ·················· 4
    1.1.5 什么是状态 ····················· 4
    1.1.6 分层的框架 ····················· 5
1.2 开发环境搭建 ·························· 5
    1.2.1 Windows 环境搭建 ············ 5
    1.2.2 MacOS 环境搭建 ············· 11
1.3 第一个 Flutter 程序 ················· 17

第2章 Flutter基础知识 ················ 23

2.1 入口程序 ······························ 23
2.2 Material Design 设计风格 ········· 24
2.3 Flutter 主题 ·························· 24
    2.3.1 创建应用主题 ················ 24
    2.3.2 局部主题 ······················ 26
    2.3.3 使用主题 ······················ 27
2.4 无状态组件和有状态组件 ········· 28
2.5 使用包资源 ··························· 31
2.6 Http 请求 ····························· 34

第3章 Dart语言简述 ···················· 40

3.1 Dart 重要概念与常用开发库 ······ 40
3.2 变量与基本数据类型 ··············· 43
3.3 函数 ····································· 46
3.4 运算符 ································· 47
3.5 流程控制语句 ························ 51
3.6 异常处理 ······························ 54
3.7 面向对象 ······························ 55
    3.7.1 实例化成员变量 ············· 55
    3.7.2 构造函数 ······················ 56
    3.7.3 读取和写入对象 ············· 57
    3.7.4 重载操作 ······················ 58
    3.7.5 继承类 ························· 59
    3.7.6 抽象类 ························· 60
    3.7.7 枚举类型 ······················ 62
    3.7.8 Mixins ························· 62
3.8 泛型 ····································· 63
3.9 库的使用 ······························ 64

3.10	异步支持	65
3.11	元数据	65
3.12	注释	67

第4章 常用组件 … 68

4.1	容器组件	68
4.2	图片组件	70
4.3	文本组件	72
4.4	图标及按钮组件	74
	4.4.1 图标组件	74
	4.4.2 图标按钮组件	75
	4.4.3 凸起按钮组件	77
4.5	列表组件	78
	4.5.1 基础列表组件	78
	4.5.2 水平列表组件	80
	4.5.3 长列表组件	82
	4.5.4 网格列表组件	83
4.6	表单组件	84

第5章 Material Design风格组件 … 88

5.1	App 结构和导航组件	89
	5.1.1 MaterialApp（应用组件）	89
	5.1.2 Scaffold（脚手架组件）	94
	5.1.3 AppBar（应用按钮组件）	95
	5.1.4 BottomNavigationBar（底部导航条组件）	97
	5.1.5 TabBar（水平选项卡及视图组件）	99
	5.1.6 Drawer（抽屉组件）	104
5.2	按钮和提示组件	107
	5.2.1 FloatingActionButton（悬停按钮组件）	107

	5.2.2 FlatButton（扁平按钮组件）	109
	5.2.3 PopupMenuButton（弹出菜单组件）	110
	5.2.4 SimpleDialog（简单对话框组件）	112
	5.2.5 AlertDialog（提示对话框组件）	113
	5.2.6 SnackBar（轻量提示组件）	115
5.3	其他组件	116
	5.3.1 TextField（文本框组件）	117
	5.3.2 Card（卡片组件）	119

第6章 Cupertino风格组件 … 122

6.1	CupertinoActivityIndicator 组件	122
6.2	CupertinoAlertDialog 对话框组件	123
6.3	CupertinoButton 按钮组件	124
6.4	Cupertino 导航组件集	125

第7章 页面布局 … 132

7.1	基础布局处理	133
	7.1.1 Container（容器布局）	133
	7.1.2 Center（居中布局）	137
	7.1.3 Padding（填充布局）	138
	7.1.4 Align（对齐布局）	140
	7.1.5 Row（水平布局）	143
	7.1.6 Column（垂直布局）	144
	7.1.7 FittedBox（缩放布局）	146
	7.1.8 Stack/Alignment	149
	7.1.9 Stack/Positioned	151
	7.1.10 IndexedStack	153

	7.1.11	OverflowBox（溢出父容器显示）……155
7.2	宽高尺寸处理……156	
	7.2.1	SizedBox（设置具体尺寸）…156
	7.2.2	ConstrainedBox（限定最大最小宽高布局）……158
	7.2.3	LimitedBox（限定最大宽高布局）……159
	7.2.4	AspectRatio（调整宽高比）…160
	7.2.5	FractionallySizedBox（百分比布局）……162
7.3	列表及表格布局……163	
	7.3.1	ListView……164
	7.3.2	GridView……166
	7.3.3	Table……167
7.4	其他布局处理……169	
	7.4.1	Transform（矩阵转换）……169
	7.4.2	Baseline（基准线布局）……171
	7.4.3	Offstage（控制是否显示组件）……172
	7.4.4	Wrap（按宽高自动换行布局）……174
7.5	布局综合示例……177	
	7.5.1	布局分析……177
	7.5.2	准备素材……179
	7.5.3	编写代码……180

第8章 手势……185

8.1 用 GestureDetector 进行手势检测……185

8.2 用 Dismissible 实现滑动删除……187

第9章 资源和图片……190

9.1 添加资源和图片……190
- 9.1.1 指定 assets……190
- 9.1.2 加载 assets……191
- 9.1.3 平台 assets……193

9.2 自定义字体……195

第10章 路由及导航……198

10.1 页面跳转基本使用……198

10.2 页面跳转发送数据……201

10.3 页面跳转返回数据……204

第11章 组件装饰和视觉效果……208

11.1 Opacity（透明度处理）……208

11.2 DecoratedBox（装饰盒子）……210

11.3 RotatedBox（旋转盒子）……217

11.4 Clip（剪裁处理）……217

11.5 案例——自定义画板……222

第12章 动画……241

12.1 用 AnimatedOpacity 实现渐变效果……241

12.2 用 Hero 实现页面切换动画……243

第13章 Flutter插件开发……246

13.1 新建插件……246

13.2 运行插件……249

13.3 示例代码分析……250

第14章 开发工具及使用技巧……259

14.1 IDE 集成开发环境……259

| | 14.1.1 Android Studio / IntelliJ ……… 259
| | 14.1.2 Visual Studio Code ………… 267
| 14.2 Flutter SDK ………………………… 274
| 14.3 使用热重载 ……………………… 275
| 14.4 格式化代码 ……………………… 276
| 14.5 Flutter 组件检查器 ……………… 278

第15章 测试与发布应用 ……………… 281

| 15.1 测试应用 ………………………… 281
| | 15.1.1 简介 ………………………… 281
| | 15.1.2 单元测试 …………………… 282
| | 15.1.3 Widget 测试 ……………… 283
| | 15.1.4 集成测试 …………………… 284
| 15.2 发布 Android 版 App …………… 286
| | 15.2.1 检查 App Manifest ………… 287
| | 15.2.2 查看构建配置 ……………… 287
| | 15.2.3 添加启动图标 ……………… 288
| | 15.2.4 App 签名 …………………… 290
| | 15.2.5 构建发布版 APK 并安装在设备上 ………………………… 291
| 15.3 发布 iOS 版 App ………………… 291
| | 15.3.1 准备工作 …………………… 291
| | 15.3.2 在 iTunes Connect 上注册应用程序 …………………… 292
| | 15.3.3 注册一个 Bundle ID ……… 292
| | 15.3.4 在 iTunes Connect 上创建应用程序记录 ………………… 293
| | 15.3.5 查看 Xcode 项目设置 …… 294
| | 15.3.6 添加应用程序图标 ………… 295
| | 15.3.7 准备发布版本 ……………… 297
| | 15.3.8 将应用发布到 App Store … 300

第16章 综合案例——即时通讯App 界面实现 …………………………… 301

| 16.1 项目介绍 ………………………… 301
| 16.2 项目搭建 ………………………… 302
| | 16.2.1 新建项目 …………………… 302
| | 16.2.2 添加源码目录及文件 ……… 305
| 16.3 入口程序 ………………………… 306
| 16.4 加载页面 ………………………… 307
| 16.5 应用页面 ………………………… 309
| 16.6 搜索页面 ………………………… 316
| | 16.6.1 布局拆分 …………………… 316
| | 16.6.2 请求获取焦点 ……………… 316
| | 16.6.3 自定义 TouchCallBack 组件 …………………………… 316
| | 16.6.4 返回文本组件 ……………… 318
| | 16.6.5 组装实现搜索页面 ………… 318
| 16.7 聊天页面 ………………………… 321
| | 16.7.1 准备聊天消息数据 ………… 321
| | 16.7.2 聊天消息列表项实现 ……… 322
| | 16.7.3 聊天消息列表实现 ………… 325
| 16.8 好友页面 ………………………… 325
| | 16.8.1 准备好友列表数据 ………… 326
| | 16.8.2 好友列表项实现 …………… 327
| | 16.8.3 好友列表头实现 …………… 329
| | 16.8.4 ContactSiderList 类 ……… 329
| | 16.8.5 Contacts 类 ………………… 332
| 16.9 我的页面 ………………………… 333
| | 16.9.1 通用列表项实现 …………… 334
| | 16.9.2 Personal 类 ………………… 335

第 1 章　开启 Flutter 之旅

Flutter 是谷歌的移动 UI 框架，可以快速在 iOS 和 Android 上构建高质量的原生用户界面。Flutter 可以与现有的代码一起工作。在全世界，Flutter 正在被越来越多的开发者和组织使用，并且 Flutter 是完全免费、开源的。

在本章中，我们将介绍 Flutter 的一些基本概念，及开发 Flutter 的准备工作，包括：
- Flutter 的特点与核心概念。
- 开发环境搭建。
- 第一个 Flutter 程序。

1.1 Flutter 的特点与核心概念

Flutter 的特点如下所示：
- **跨平台**：现在 Flutter 至少可以跨 5 种平台，甚至支持嵌入式开发。我们常用的有 MacOS、Windows、Linux、Android、iOS，甚至可以在谷歌最新的操作系统 Fuchsia 上运行。到目前为止，Flutter 算是支持平台最多的框架了，良好的跨平台性，直接带来的好处就是减少开发成本。
- **丝滑般的体验**：使用 Flutter 内置高大上的 Material Design 和 Cupertino 风格组件、丰富的 motion API、平滑而自然的滑动效果和平台感知，为用户带来全新体验。
- **响应式框架**：使用 Flutter 的、响应式框架和一系列基础组件，可以轻松构建用户界面。使用功能强大且灵活的 API（针对 2D、动画、手势、效果等）能解决艰难的 UI 挑战。

- **支持插件**：通过 Flutter 的插件可以访问平台本地 API，如相机、蓝牙、WiFi 等。借助现有的 Java、Swift、Objective C、C++ 代码实现对原生系统的调用。
- **60fps 超高性能**：Flutter 采用 GPU 渲染技术，所以性能极高。Flutter 编写的应用是可以达到 60fps（每秒传输帧数），这也就是说，它完全可以胜任游戏的制作。官方宣称用 Flutter 开发的应用甚至会超过原生应用的性能。

Flutter 包括一个现代的响应式框架、一个 2D 渲染引擎、现成的组件和开发工具。这些组件可以帮助你快速地设计、构建、测试和调试应用程序。Flutter 的核心概念有：组件、构建、状态、框架等，下面分别介绍。

1.1.1 一切皆为组件

组件（Widget）是 Flutter 应用程序用户界面的基本构建块。不仅按钮、输入框、卡片、列表这些内容可作为 Widget，甚至将布局方式、动画处理都视为 Widget。所以 Flutter 具有一致的统一对象模型：Widget。

Widget 可以定义为：
- 一个界面组件（如按钮或输入框）。
- 一个文本样式（如字体或颜色）。
- 一种布局（如填充或滚动）。
- 一种动画处理（如缓动）。
- 一种手势处理（GestureDetector）。

Widget 具有丰富的**属性**及**方法**，属性通常用来改变组件的状态（颜色、大小等）及回调方法的处理（单击事件回调、手势事件回调等）。方法主要是提供一些组件的功能扩展。比如：TextBox 是一个矩形的文本组件，其属性及方法如下：

- bottom：底部间距属性。
- direction：文本排列方向属性。
- left：左侧间距属性。
- right：右侧间距属性。
- top：上部间距属性。
- toRect：导出矩形方法。
- toString：转换成字符串方法。

1.1.2 组件嵌套

复杂的功能界面通常都是由一个一个简单功能的组件组装完成的。有的组件负责布局，有的负责定位，有的负责调整大小，有的负责渐变处理，等等。这种嵌套组合的方式带来的最大好处就是解耦。

例如，界面中添加了一个居中组件 Center，居中组件里嵌套了一个容器组件 Container，容器组件里嵌套了一个文本组件 Text 和一个装饰器 BoxDecoration。代码如下所示：

```
return new Center(
  // 添加容器
  child: new Container(
    // 添加装饰器
    decoration: new BoxDecoration(
    ),
    child: new Text(
      // 添加文本组件
    ),
  ),
),
```

大家如果是首次看到这段代码会觉得嵌套层次太多，太复杂。其实不然，随着对组件的深入了解及熟练使用，写起来还是非常得心应手的。

最基础的组件类是 Widget，其他所有的组件都是继承 Widget 的，如图 1-1 所示。紧接着下面有两大类组件：**有状态组件**及**无状态组件**。有状态组件是界面会发生变化的组件，如 Scrollable、Animatable 等，无状态的组件即界面不发生变化的组件，如 Text、AssetImage 等。

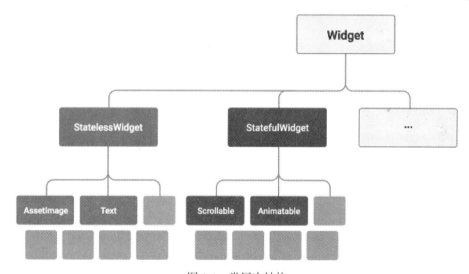

图 1-1 类层次结构

1.1.3 构建 Widget

可以重写 Widget 的 build 方法来构建一个组件，如下代码所示：

```
@protected Widget build(BuildContext context);
```

build 即为创建一个 Widget 的意思，返回值也是一个 Widget 对象，不管返回的是单个组件还是返回通过嵌套的方式组合的组件，都是 Widget 的实例。

1.1.4 处理用户交互

如果 Widget 需要根据用户交互或其他因素进行更改,则该 Widget 是有状态的。例如,如果一个 Widget 的计数器在用户点击一个按钮时递增,那么该计数器的值就是该 Widget 的状态。当该值发生变化时,需要重新构建 Widget 以更新 UI。

这些 Widget 将继承 StatefulWidget(而不是 State)并将它们的可变状态存储在 State 的子类中,如图 1-2 所示。

每当你改变一个 State 对象时(例如增加计数器),必须调用 setState() 来通知框架,框架会再次调用 State 的构建方法来更新用户界面。

有了独立的状态和 Widget 对象,其他 Widget 可以以同样的方式处理无状态和有状态的 Widget,而不必担心丢失状态。父 Widget 可以自由地创造子 Widget 的新实例且不会失去子 Widget 的状态,而不是通过持有子 Widget 来维持其状态。框架在适当的时候完成查找和重用现有状态对象的所有工作。

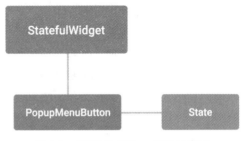

图 1-2 有状态 Widget 继承示意图

1.1.5 什么是状态

Flutter 中的状态和 React 中的状态概念一致。React 的核心思想是组件化的思想,应用由组件搭建而成,而组件中最重要的概念是 State(状态),State 是一个组件的 UI 数据模型,是组件渲染时的数据依据。Flutter 程序的运行可以认为是一个巨大的状态机,用户的操作、请求 API 和系统事件的触发都是推动状态机运行的触发点,触发点通过调用 setState 方法推动状态机进行响应。状态的生命周期如图 1-3 所示。

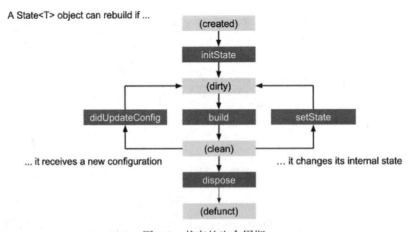

图 1-3 状态的生命周期

1.1.6 分层的框架

Flutter 框架是一个分层的结构,每一层都建立在前一层之上。图 1-4 显示了 Flutter 框架,上层比下层的使用频率更高。

> **提示** 有关构成 Flutter 分层框架的完整库,请参阅官方的 API 文档,地地为:https://docs.flutter.io/。

分层设计的目标是帮助开发者用更少的代码做更多的事情。例如,Material 层通常组合来自 Widget 层的基本 Widget,而 Widget 层通过较低级对象渲染层来构建。

分层结构为构建应用程序提供了许多选项。选择一种自定义的方法来释放框架的全部表现力,或者使用构件层中的构建块,或混合搭配。可以使用 Flutter 提供的所有现成的 Widget,也可以使用 Flutter 团队用于构建框架的相同工具和技术创建定制的 Widget。也就是说,你可以从高层次、统一的 Widget 概念中获得开发效率优势,也可以深入到下层施展才能。

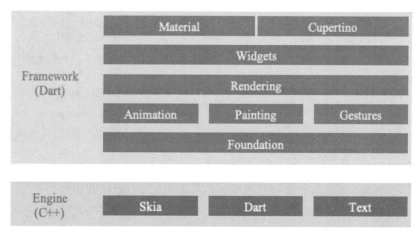

图 1-4　Flutter 框架

1.2　开发环境搭建

开发环境搭建还是非常烦琐的,任何一个步骤失败都会导致最终环境搭建不能完成。Flutter 支持三种环境:Windows、MacOS 和 Linux。这里我们主要讲解 Windows 及 MacOS 的环境搭建。

1.2.1　Windows 环境搭建

1. 使用镜像

首先解决网络问题。环境搭建过程中需要下载很多资源文件,当某个资源更新不到时,就可能会报各种错误。在国内访问 Flutter 有时可能会受到限制，Flutter 官方为中国开发者搭建了临时镜像，大家可以将如下环境变量加入到用户环境变量中：

```
export PUB_HOSTED_URL=https://pub.flutter-io.cn
export FLUTTER_STORAGE_BASE_URL=https://storage.flutter-io.cn
```

 此镜像为临时镜像，并不能保证一直可用，读者可以参考 Using Flutter in China：https://github.com/flutter/flutter/wiki/Using-Flutter-in-China 以获得有关镜像服务器的最新动态。

2. 安装 Git

Flutter 依赖的命令行工具为 Git for Windows（Git 命令行工具）。Windows 版本的下载地址为：https://git-scm.com/download/win。

3. 下载安装 Flutter SDK

去 Flutter 官网下载其最新可用的安装包。

 Flutter 的渠道版本会不停变动，请以 Flutter 官网为准。Flutter 官网下载地址：https://flutter.io/docs/development/tools/sdk/archive#windows。Flutter GitHub 下载地址：https://github.com/flutter/flutter/releases。

将安装包 zip 解压到你想安装 Flutter SDK 的路径（如：D:\Flutter）。在 Flutter 安装目录的 Flutter 文件下找到 flutter_console.bat，双击运行并启动 Flutter 命令行，接下来，你就可以在 Flutter 命令行运行 flutter 命令了。

 不要将 Flutter 安装到需要一些高权限的路径如 C:\Program Files\。

4. 添加环境变量

不管使用什么工具，如果想在系统的任意地方能够运行这个工具的命令，则需要添加工具的路径到系统 path 里去。这里路径指向到 Flutter 文件的 bin 路径，如图 1-5 所示。同时，检查是否有名为 "PUB_HOSTED_URL" 和 "FLUTTER_STORAGE_BASE_URL" 的条目，如果没有，也需要添加它们。重启 Windows 才能使更改生效。

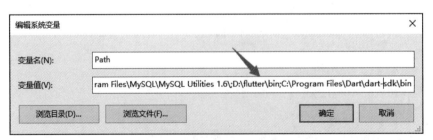

图 1-5　添加 Flutter 环境变量

5. 运行 flutter 命令安装各种依赖

使用 Windows 命令窗口运行以下命令，查看是否需要安装任何依赖项来完成安装：

```
flutter doctor
```

该命令检查你的环境并在终端窗口中显示报告。Dart SDK 已经捆绑在 Flutter 里了，没有必要单独安装 Dart。 仔细检查命令行输出以获取可能需要安装的其他软件或进一步需要执行的任务。如下代码粗体显示，Android SDK 缺少命令行工具，需要下载并且提供了下载地址，通常这种情况只需要把网络连好，VPN 开好，然后重新运行 flutter doctor 命令即可。

```
[-] Android toolchain - develop for Android devices
  · Android SDK at D:\Android\sdk
  × Android SDK is missing command line tools; download from https://goo.gl/
    XxQghQ
  · Try re-installing or updating your Android SDK,
    visit https://flutter.io/setup/#android-setup for detailed instructions.
```

注意 一旦你安装了任何缺失的依赖，需再次运行 flutter doctor 命令来验证你是否已经正确地设置了，同时需要检查移动设备是否连接正常。

6. 编辑器设置

如果使用 flutter 命令行工具，可以使用任何编辑器来开发 Flutter 应用程序。输入 flutter help 在提示符下查看可用的工具。但是笔者建议最好安装一款功能强大的 IDE 来进行开发，毕竟开发调试运行打包的效率会更高。由于 Windows 环境只能开发 Flutter 的 Android 应用，所以接下来我们会重点介绍 Android Studio 这款 IDE。

（1）安装 Android Studio

要为 Android 开发 Flutter 应用，你可以使用 Mac 或 Windows 操作系统。Flutter 需要安装和配置 Android Studio，步骤如下：

1）下载并安装 Android Studio：https://developer.android.com/studio/index.html。

2）启动 Android Studio，然后执行"Android Studio 安装向导"。这将安装最新的 Android SDK、Android SDK 平台工具和 Android SDK 构建工具，这是 Flutter 为 Android 开发时所必需的。

（2）设置你的 Android 设备

要准备在 Android 设备上运行并测试你的 Flutter 应用，需要安装 Android 4.1（API level 16）或更高版本的 Android 设备。步骤如下：

1）在你的设备上启用"开发人员选项"和"USB 调试"，这些选项通常在设备的"设置"界面里。

2）使用 USB 线将手机与计算机连接。如果你的设备出现提示，请授权计算机访问你的设备。

3）在终端中，运行 flutter devices 命令以验证 Flutter 识别出你连接的 Android 设备。

4）用 flutter run 启动你的应用程序。

> **提示** 默认情况下，Flutter 使用的 Android SDK 版本是基于你的 adb 工具版本。如果你想让 Flutter 使用不同版本的 Android SDK，则必须将该 ANDROID_HOME 环境变量设置为 SDK 安装目录。

（3）设置 Android 模拟器

要准备在 Android 模拟器上运行并测试 Flutter 应用，请按照以下步骤操作：

1）启动 Android Studio → Tools → Android → AVD Manager 并选择 Create Virtual Device，打开虚拟设备面板，如图 1-6 所示。

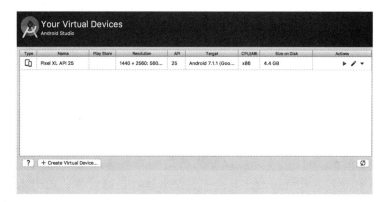

图 1-6　打开虚拟设备面板

2）选择一个设备并点击 Next，如图 1-7 所示。

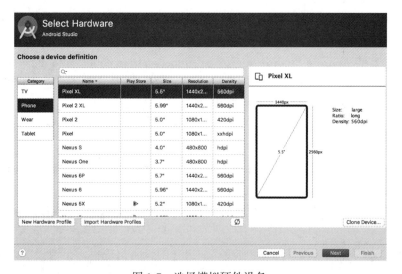

图 1-7　选择模拟硬件设备

3）选择一个镜像点击 download 即可，然后点击 Next，如图 1-8 所示。

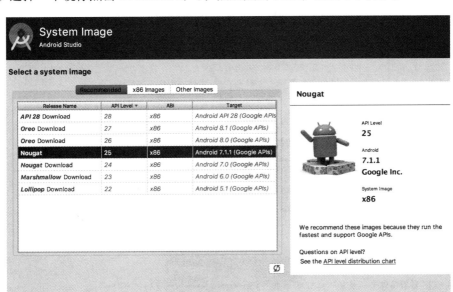

图 1-8　选择系统镜像

4）验证配置信息，填写虚拟设备名称，选择 Hardware - GLES 2.0 以启用硬件加速，点击 Finish，如图 1-9 所示。

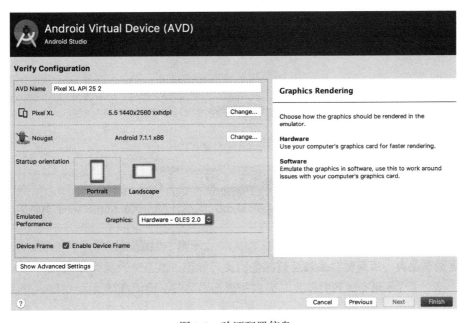

图 1-9　验证配置信息

5）在工具栏选择刚刚添加的模拟器，如图1-10所示。

图1-10　在工具栏选择模拟器

6）也可以在命令行窗口运行flutter run启动模拟器。当能正常显示模拟器时（如图1-11所示），则表示模拟器安装正常。

图1-11　Android模拟器运行效果图

> 提示　建议选择当前主流手机型号作为模拟器，开启硬件加速，使用x86或x86_64 image。详细文档请参考：https://developer.android.com/studio/run/emulator-acceleration.html。

（4）安装Flutter和Dart插件

IDE需要安装两个插件：

❏ Flutter插件：支持Flutter开发工作流（运行、调试、热重载等）。

❏ Dart插件：提供代码分析（输入代码时进行验证、代码补全等）。

打开Android Studio的系统设置面板，找到Plugins分别搜索Flutter和Dart，点击安装即可,如图 1-12 所示。

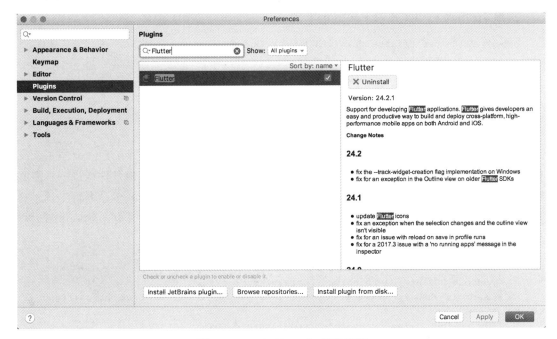

图 1-12　Android Studio 插件安装

1.2.2　MacOS 环境搭建

首先解决网络问题,参见上节"Windows 环境搭建"。

1. 命令行工具

Flutter 依赖的命令行工具有 bash、mkdir、rm、git、curl、unzip、which。

2. 下载安装 Flutter SDK

请按以下步骤进行下载安装 Flutter SDK:

步骤 1:去 Flutter 官网下载其最新可用的安装包。

> **注意** Flutter 的渠道版本会不停变动,请以 Flutter 官网为准。另外,在中国大陆地区,要想获取安装包列表或下载安装包有可能发生困难,读者也可以去 Flutter GitHub 项目下去下载安装 Release 包。
>
> Flutter 官网下载地址:https://flutter.io/docs/development/tools/sdk/archive#macos
>
> Flutter GitHub 下载地址:https://github.com/flutter/flutter/releases

步骤 2:解压安装包到你想安装的目录,如:

```
cd /Users/ksj/Desktop/flutter/
```

```
unzip /Users/ksj/Desktop/flutter/v0.11.9.zip.zip
```

步骤 3：添加 Flutter 相关工具到 path 中：

```
export PATH=`pwd`/flutter/bin:$PATH
```

3. 运行 Flutter 命令安装各种依赖

运行以下命令查看是否需要安装其他依赖项：

```
flutter doctor
```

该命令检查你的环境并在终端窗口中显示报告。Dart SDK 已经捆绑在 Flutter 里了，没有必要单独安装 Dart。仔细检查命令行输出以获取可能需要安装的其他软件或进一步需要执行的任务（以粗体显示）。如下代码中粗体提示表示，Android SDK 缺少命令行工具，需要下载并且提供了下载地址，通常这种情况只需要把网络连好，VPN 开好，然后重新运行 flutter doctor 命令。

```
[-] Android toolchain - develop for Android devices
    · Android SDK at /Users/obiwan/Library/Android/sdk
    ✗ Android SDK is missing command line tools; download from https://goo.gl/
XxQghQ
    · Try re-installing or updating your Android SDK,
      visit https://flutter.io/setup/#android-setup for detailed instructions.
```

 一旦你安装了任何缺失的依赖，需再次运行 flutter doctor 命令来验证你是否已经正确地设置了，同时需要检查移动设备是否连接正常。

4. 添加环境变量

使用 vim 命令打开～/.bash_profile 文件，添加如下内容：

```
export ANDROID_HOME=~/Library/Android/sdk //android sdk 目录
export PATH=$PATH:$ANDROID_HOME/tools:$ANDROID_HOME/platform-tools
export PUB_HOSTED_URL=https://pub.flutter-io.cn //国内用户需要设置
export FLUTTER_STORAGE_BASE_URL=https://storage.flutter-io.cn //国内用户需要设置
export PATH=/Users/ksj/Desktop/flutter/flutter/bin:$PATH // 直接指定 flutter 的 bin
                                                                                 地址
```

 请将 PATH=/Users/ksj/Desktop/flutter/flutter/bin 更改为你的路径即可。

完整的环境变量设置如图 1-13 所示。

设置好环境变量以后，请务必运行 source $HOME/.bash_profile 刷新当前终端窗口，以使刚刚配置的内容生效。

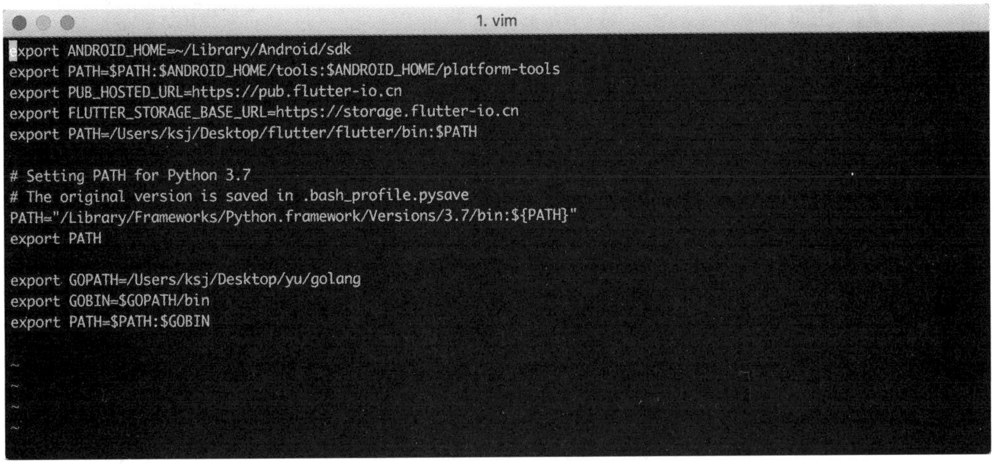

```
export ANDROID_HOME=~/Library/Android/sdk
export PATH=$PATH:$ANDROID_HOME/tools:$ANDROID_HOME/platform-tools
export PUB_HOSTED_URL=https://pub.flutter-io.cn
export FLUTTER_STORAGE_BASE_URL=https://storage.flutter-io.cn
export PATH=/Users/ksj/Desktop/flutter/flutter/bin:$PATH

# Setting PATH for Python 3.7
# The original version is saved in .bash_profile.pysave
PATH="/Library/Frameworks/Python.framework/Versions/3.7/bin:${PATH}"
export PATH

export GOPATH=/Users/ksj/Desktop/yu/golang
export GOBIN=$GOPATH/bin
export PATH=$PATH:$GOBIN
```

图 1-13　MacOS 环境变量设置

5. 编辑器设置

如果使用 Flutter 命令行工具，可以使用任何编辑器来开发 Flutter 应用程序。输入 flutter help 在提示符下查看可用的工具。但是笔者建议最好安装一款功能强大的 IDE 来进行开发，毕竟开发调试运行打包的效率会更高。由于 MacOS 环境既能开发 Android 应用也能开发 iOS 应用，Android 设置请参考 1.2.1 节"Windows 环境搭建"中的"安装 Android Studio"，接下来我们会介绍 Xcode 使用方法。

（1）安装 Xcode

安装最新 Xcode。通过链接下载：https://developer.apple.com/xcode/，或通过苹果应用商店下载：https://itunes.apple.com/us/app/xcode/id497799835。

（2）设置 iOS 模拟器

要准备在 iOS 模拟器上运行并测试你的 Flutter 应用。要打开一个模拟器，在 MacOS 的终端输入以下命令：

```
open -a Simulator
```

可以找到并打开默认模拟器。如果想切换模拟器，可以打开 Hardware 下在 Device 菜单选择某一个模拟器，如图 1-14 所示。

打开后的模拟器如图 1-15 所示。

接下来，在终端运行 flutter run 命令或者打开 Xcode，如图 1-16 所示选择好模拟器。点击运行按钮即可启动你的应用。

（3）安装到 iOS 设备

要在苹果真机上测试 Flutter 应用，需要一个苹果开发者账户，并且还需要在 Xcode 中进行设置。

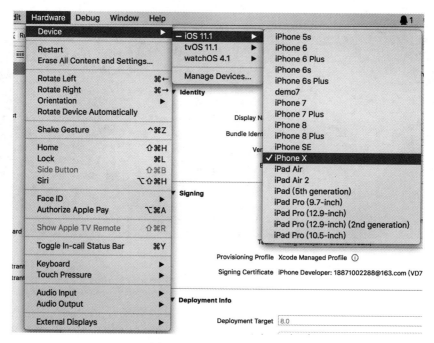

图 1-14　选择 iOS 模拟器

图 1-15　iOS 模拟器效果图

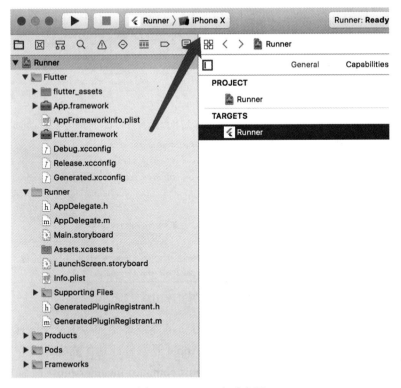

图 1-16　Xcode 启动应用

1）安装 Homebrew 工具，Homebrew 是一款 MacOS 平台下的软件包管理工具，拥有安装、卸载、更新、查看、搜索等很多实用的功能。下载地址为：https://brew.sh。

2）打开终端并运行一些命令，安装用于将 Flutter 应用安装到 iOS 设备的工具，命令如下所示：

```
brew update
brew install --HEAD libimobiledevice
brew install ideviceinstaller ios-deploy cocoapods
pod setup
```

> 提示　如果这些命令中有任何一个失败并出现错误，请运行 brew doctor 并按照说明解决问题。

接下来需要 Xcode 签名。Xcode 签名设置有以下几个步骤：

步骤 1：在你 Flutter 项目目录中通过双击 ios/Runner.xcworkspace 打开默认的 Xcode 工程。

步骤 2：在 Xcode 中，选择导航面板左侧中的 Runner 项目。

步骤 3：在 Runner target 设置页面中，确保在 General → Signing → Team（常规→签名→

团队）下选择了你的开发团队，如图 1-17 所示。当你选择一个团队时，Xcode 会创建并下载开发证书，为你的设备注册你的账户，并创建和下载配置文件。

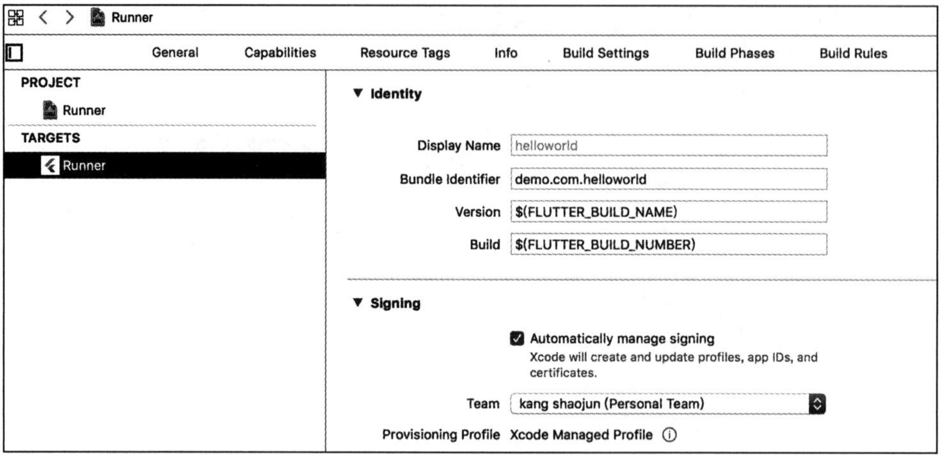

图 1-17　设置开发团队

步骤 4：要开始你的第一个 iOS 开发项目，可能需要使用你的 Apple ID 登录 Xcode。任何 Apple ID 都支持开发和测试。需要注册 Apple 开发者计划才能将你的应用分发到 App Store。请查看 https://developer.apple.com/support/compare-memberships/ 这篇文章。登录界面如图 1-18 所示。

图 1-18　使用 Apple ID

步骤 5：当你第一次添加真机设备进行 iOS 开发时，需要同时信任你的 Mac 和该设备上的开发证书。点击 Trust 即可，如图 1-19 所示。

步骤 6：如果 Xcode 中的自动签名失败，请查看项目的 Bundle Identifier 值是否唯一。这个 ID 即为应用的唯一 ID，建议使用域名反过来写，如图 1-20 所示。

步骤 7：使用 flutter run 命令运行应用程序。

图 1-19　信任此电脑图示

图 1-20　验证 Bundle Identifier 值

1.3　第一个 Flutter 程序

万事开头难，我们用 Hello World 来看一个最简单的 Flutter 工程，具体步骤如下。

步骤 1：新建一个 Flutter 工程，选择 Flutter Application，如图 1-21 所示。

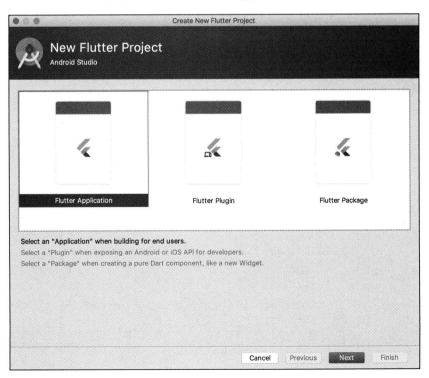

图 1-21　新建工程

步骤 2：点击 Next 按钮，打开应用配置界面，其中在 Project name 中填写 helloworld，Flutter SDK path 使用默认值，IDE 会根据 SDK 安装路径自动填写，Project location 填写为工程放置的目录，在 Description 中填写项目描述，任意字符即可，如图 1-22 所示。

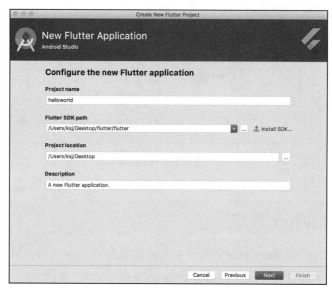

图 1-22　配置 Flutter 工程

步骤 3：点击 Next 按钮，打开包设置界面，在 Company domain 中填写域名，注意域名要反过来写，这样可以保证全球唯一，Platform channel language 下面的两个选项不需要勾选，如图 1-23 所示。

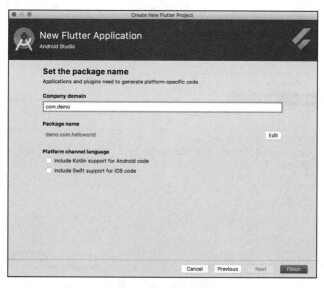

图 1-23　设置包名界面

步骤 4：点击 Finish 按钮开始创建第一个工程，等待几分钟，会创建如图 1-24 所示工程。

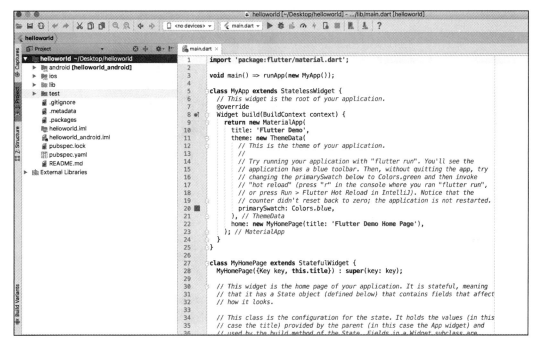

图 1-24　示例工程主界面

步骤 5：工程建好后，可以先运行一下看看根据官方创建的示例运行的效果，点击 Open iOS Simulator 打开 iOS 模拟器，具体操作如图 1-25 所示。

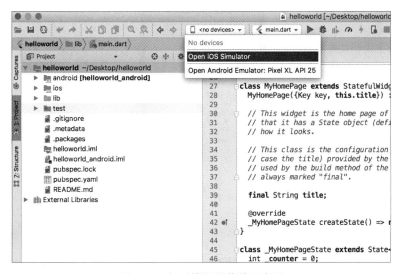

图 1-25　打开模拟器菜单示意图

步骤 6：等待几秒钟后会打开模拟器，如图 1-26 所示。

图 1-26　模拟器启动完成图

步骤 7：点击 debug(调试) 按钮，启动官方示例程序，点击 + 号按钮，可以自动加 1，此示例是一个基于 Material Design 风格的应用程序，如图 1-27 所示。

图 1-27　官方示例运行效果图

步骤 8：接下来我们打开工程目录下的 main.dart 文件，清空 main.dart 代码，如图 1-28 箭头所指。

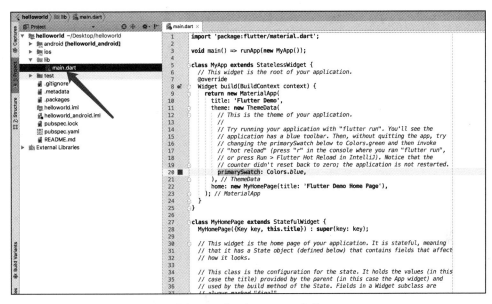

图 1-28　打开 main.dart 文件

步骤 9：把 Hello World 代码粘贴至 main.dart 文件里，完整代码如下所示：

```dart
import 'package:flutter/material.dart';

void main() => runApp(MyApp());

class MyApp extends StatelessWidget {
  @override
  Widget build(BuildContext context) {
    return MaterialApp(
      title: 'Welcome to Flutter',
      home: Scaffold(
        appBar: AppBar(
          title: Text('Welcome to Flutter'),
        ),
        body: Center(
          child: Text('Hello World'),
        ),
      ),
    );
  }
}
```

步骤 10：重新运行此程序，标题栏显示 Welcome to Flutter，页面中间显示 Hello World。这样，第一个 Flutter 程序就运行出来啦，如图 1-29 所示。

图 1-29　Hello World 运行效果图

第 2 章 Chapter 2

Flutter 基础知识

在创建了第一个 Flutter 程序后，我们还需要继续补充 Flutter 的基础知识。在后面的章节中讲解组件、布局、动画、装饰等时都需要用到这些基础知识。

本章将围绕以下几个知识点展开：

- 入口程序
- Material Design
- Flutter 主题
- 无状态组件和有状态组件
- 使用包资源
- Http 请求

2.1 入口程序

每一个 Flutter 项目的 /lib 目录下都有一个 main.dart 文件，打开该文件，里面应该有一个 main() 函数。Flutter 使用 Dart 语言开发，而在 Dart 语言中，main() 函数是 Dart 程序的入口，也就是说，Flutter 程序在运行的时候，第一个执行的函数就是 main() 函数。如下面的代码所示：

```
void main() => runApp(Widget app);
```

如果你是第一次接触 Dart 语言，可能会对上面的语法感到陌生，这是 Dart 语言特有的速写形式，将其展开后，完整代码如下所示：

```
void main() {
  return runApp(Widget app);
}
```

从上面的代码中可以看到，main() 函数中只调用 runApp 函数，使用 runApp 函数可以将给定的根组件填满整个屏幕。你可能会有疑问，为什么一定要使用 runApp 函数？如果不调用 runApp 函数，项目也可以正常执行，但是屏幕上什么都不会显示。Flutter 是 Dart 语言的移动应用框架，runApp 函数就是 Flutter 框架的入口，如果不调用 runApp 函数，那你执行的就是一个 Dart 控制台应用。更多关于 Dart 语言的细节，会在下面第 3 章 "Dart 语言简述"专门讲解。

2.2 Material Design 设计风格

每一个 .dart 文件的第一行几乎都会导入 flutter/material.dart 包，这个包是 Flutter 实现 Material Design 设计风格的基础包，里面有文本输入框（Text）、图标（Icon）、图片（Image）、行排列布局（Row）、列排列布局（Column）、Decoration（装饰器）、动画等组件，大家可以将它们理解为网页中的按钮、标题、选项框等组件库。第一行代码如下所示：

```
import 'package:flutter/material.dart';
```

那么 Material Design 又是什么呢？是谷歌推出的一套视觉设计语言。比如有的 App 可以换皮肤，而每一套皮肤就是一种设计语言，有古典风、炫酷风、极简风，等等，而 Material Design 就是谷歌风。Flutter 采用的就是 Material Design 风格。

2.3 Flutter 主题

为了在整个应用中使用同一套颜色和字体样式，可以使用"主题"这种方式。定义主题有两种方式：全局主题，或使用 Theme 来定义应用程序局部的颜色和字体样式。事实上，全局主题只是由应用程序根 MaterialApp 创建的主题（Theme）。

定义一个主题后，就可以在我们自己的 Widget 中使用它，Flutter 提供的 Material Widgets 将使用主题为 AppBars、Buttons、Checkboxes 等设置背景颜色和字体样式。

2.3.1 创建应用主题

创建主题的方法是将 ThemeData 提供给 MaterialApp 构造函数，这样就可以在整个应用程序中共享包含颜色和字体样式的主题。ThemeData 的主要属性如表 2-1 所示。

表 2-1 ThemeData 属性及描述

属性名	类型	说明
accentColor	Color	前景色（文本、按钮等）
accentColorBrightness	Brightness	accentColor 的亮度。用于确定放置在突出颜色顶部的文本和图标的颜色（例如 FloatingButton 上的图标）
accentIconTheme	IconThemeData	与突出颜色对比的图标主题
accentTextTheme	TextTheme	与突出颜色对比的文本主题
backgroundColor	Color	与 primaryColor 对比的颜色（例如：用作进度条的剩余部分）
bottomAppBarColor	Color	BottomAppBar 的默认颜色
brightness	Brightness	应用程序整体主题的亮度。由按钮等 Widget 使用，以确定在不使用主色或强调色时要选择的颜色
buttonColor	Color	Material 中 RaisedButtons 使用的默认填充色
buttonTheme	ButtonThemeData	定义了按钮等控件的默认配置，如 RaisedButton 和 FlatButton
canvasColor	Color	MaterialType.canvas Material 的默认颜色
cardColor	Color	Material 被用作 Card 时的颜色
chipTheme	ChipThemeData	用于渲染 Chip 的颜色和样式
dialogBackgroundColor	Color	Dialog 元素的背景色
disabledColor	Color	用于 Widget 无效的颜色，包括任何状态。例如禁用复选框
dividerColor	Color	Dividers 和 PopupMenuDividers 的颜色，也用于 ListTiles 中间和 DataTables 的每行中间
errorColor	Color	用于输入验证错误的颜色，例如在 TextField 中
hashCode	int	对象的哈希值
highlightColor	Color	用于类似墨水喷溅动画或指示菜单被选中的高亮颜色
iconTheme	IconThemeData	与卡片和画布颜色形成对比的图标主题
indicatorColor	Color	TabBar 中选项选中的指示器颜色
inputDecorationTheme	InputDecorationTheme	InputDecorator、TextField 和 TextFormField 的默认 InputDecoration 值基于此主题
platform	TargetPlatform	Widget 需要适配的目标类型
primaryColor	Color	App 主要部分的背景色（ToolBar、Tabbar 等）
primaryColorBrightness	Brightness	primaryColor 的亮度
primaryColorDark	Color	primaryColor 的较暗版本
primaryColorLight	Color	primaryColor 的较亮版本
primaryIconTheme	IconThemeData	一个与主色对比的图标主题
primaryTextTheme	TextThemeData	一个与主色对比的文本主题
scaffoldBackgroundColor	Color	作为 Scaffold 基础的 Material 默认颜色，典型 Material 应用或应用内页面的背景颜色
secondaryHeaderColor	Color	有选定行时 PaginatedDataTable 标题的颜色
selectedRowColor	Color	选中行时的高亮颜色
sliderTheme	SliderThemeData	用于渲染 Slider 的颜色和形状

(续)

属性名	类型	说明
splashColor	Color	墨水喷溅的颜色
splashFactory	InteractiveInkFeatureFactory	定义 InkWall 和 InkResponse 生成的墨水喷溅的外观
textSelectionColor	Color	文本字段中选中文本的颜色，例如 TextField
textSelectionHandleColor	Color	用于调整当前文本的哪个部分的句柄颜色
textTheme	TextTheme	与卡片和画布对比的文本颜色
toggleableActiveColor	Color	用于突出显示切换 Widget（如 Switch、Radio 和 Checkbox）的活动状态的颜色
unselectedWidgetColor	Color	用于 Widget 处于非活动（但已启用）状态的颜色。例如，未选中的复选框。通常与 accentColor 形成对比
runtimeType	Type	表示对象的运行时类型

如果没有提供主题，Flutter 将创建一个默认主题。主题数据的示例代码如下：

```
new MaterialApp(
  title: title,
  theme: new ThemeData(
    brightness: Brightness.dark,
    primaryColor: Colors.lightBlue[800],
    accentColor: Colors.cyan[600],
  ),
);
```

2.3.2 局部主题

如果我们想在应用程序的某一部分使用特殊的颜色，那么就需要覆盖全局的主题。有两种方法可以解决这个问题：创建特有的主题数据或扩展父主题。

1. 创建特有的主题数据

实例化一个 ThemeData 并将其传递给 Theme 对象，代码如下：

```
new Theme(
  // 创建一个特有的主题数据
  data: new ThemeData(
    accentColor: Colors.yellow,
  ),
  child: new FloatingActionButton(
    onPressed: () {},
    child: new Icon(Icons.add),
  ),
);
```

2. 扩展父主题

扩展父主题时无须覆盖所有的主题属性，我们可以通过使用 copyWith 方法来实现，代码如下：

```
new Theme(
    // 覆盖accentColor为Colors.yellow
    data: Theme.of(context).copyWith(accentColor: Colors.yellow),
    child: new FloatingActionButton(
        onPressed: null,
        child: new Icon(Icons.add),
    ),
);
```

2.3.3　使用主题

主题定义好后就可以使用它了。首先，函数Theme.of(context)可以通过上下文来获取主题，方法是查找最近的主题，如果找不到就会找整个应用的主题。

下面来看一个简单的示例，应用的主题颜色定义为绿色，界面中间再加一个带有背景色的文本。

完整的例子代码如下所示：

```
import 'package:flutter/foundation.dart';
import 'package:flutter/material.dart';

void main() {
  runApp(new MyApp());
}

class MyApp extends StatelessWidget {
  @override
  Widget build(BuildContext context) {
    final appName = '自定义主题';

    return new MaterialApp(
      title: appName,
      theme: new ThemeData(
        brightness: Brightness.light,// 应用程序整体主题的亮度
        primaryColor: Colors.lightGreen[600],//App主要部分的背景色
        accentColor: Colors.orange[600],// 前景色（文本、按钮等）
      ),
      home: new MyHomePage(
        title: appName,
      ),
    );
  }
}

class MyHomePage extends StatelessWidget {
  final String title;

  MyHomePage({Key key, @required this.title}) : super(key: key);

  @override
```

```
  Widget build(BuildContext context) {
    return new Scaffold(
      appBar: new AppBar(
        title: new Text(title),
      ),
      body: new Center(
        child: new Container(
          // 获取主题的 accentColor
          color: Theme.of(context).accentColor,
          child: new Text(
            '带有背景颜色的文本组件',
            style: Theme.of(context).textTheme.title,
          ),
        ),
      ),
      floatingActionButton: new Theme(
        // 使用 copyWith 的方式获取 accentColor
        data: Theme.of(context).copyWith(accentColor: Colors.grey),
        child: new FloatingActionButton(
          onPressed: null,
          child: new Icon(Icons.computer),
        ),
      ),
    );
  }
}
```

自定义主题的效果如图 2-1 所示。

2.4 无状态组件和有状态组件

无状态组件（StatelessWidget）是不可变的，这意味着它们的属性不能改变，所有的值都是最终的。

有状态组件（StatefulWidget）持有的状态可能在 Widget 生命周期中发生变化。实现一个 StatefulWidget 至少需要两个类：

- 一个 StatefulWidget 类。
- 一个 State 类。StatefulWidget 类本身是不变的，但是 State 类在 Widget 生命周期中始终存在。

Flutter 的官方给出一个有状态组件的示例，点击右下角的 + 号按钮，应用界面中间的数字会加 1，如图 2-2 所示。

这个示例有几个关键的部分，解析如下。

示例代码中 MyHomePage 必须继承自 StatefulWidget 类，如下所示：

图 2-1　自定义主题效果图

图 2-2　Flutter 官方示例

```
class MyHomePage extends StatefulWidget
```

重写 createState 方法，如下所示：

```
@override
MyHomePageState createState() => new _MyHomePageState();
```

状态类必须继承自 State 类，如下所示：

```
class _MyHomePageState extends State<MyHomePage>
```

定义一个普通变量 _counter 作为计数器变量，调用 setState 方法来控制这个变量的值的变化，如下所示：

```
int _counter = 0;

void _incrementCounter() {
  setState(() {
    // 计数器变量
    _counter++;
  });
}
```

完整的示例代码如下所示：

```dart
import 'package:flutter/material.dart';

void main() => runApp(new MyApp());

//MyApp 不需要做状态处理，所以此组件继承 StatelessWidget 即可
class MyApp extends StatelessWidget {
  // 这个组件是整个应用的主组件
  @override
  Widget build(BuildContext context) {
    return new MaterialApp(
      title: 'Flutter Demo',
      theme: new ThemeData(
        // 自定义主题
        primarySwatch: Colors.blue,
      ),
      home: new MyHomePage(title: 'Flutter Demo Home Page'),
    );
  }
}

// 主页需要继承自 StatefulWidget
class MyHomePage extends StatefulWidget {
  MyHomePage({Key key, this.title}) : super(key: key);

  // 标题
  final String title;

  // 必须重写 createState 方法
  @override
  _MyHomePageState createState() => new _MyHomePageState();
}
// 状态类必须继承 State 类，注意后面需要指定为 <MyHomePage>
class _MyHomePageState extends State<MyHomePage> {
  int _counter = 0;// 计数器

  void _incrementCounter() {
    // 调用 State 类里的 setState 方法来更改状态值，使得计数器加 1
    setState(() {
      // 计数器变量，每次点击让其加 1
      _counter++;
    });
  }

  @override
  Widget build(BuildContext context) {

    return new Scaffold(
      appBar: new AppBar(
        title: new Text(widget.title),
      ),
```

```
      //居中布局
      body: new Center(
      
        //垂直布局
        child: new Column(
          //主轴居中对齐
          mainAxisAlignment: MainAxisAlignment.center,
          children: <Widget>[
            new Text(
              'You have pushed the button this many times:',
            ),
            new Text(
              '$_counter',//绑定计数器的值
              style: Theme.of(context).textTheme.display1,
            ),
          ],
        ),
      ),
      floatingActionButton: new FloatingActionButton(
        onPressed: _incrementCounter,//按下+号按钮调用自增函数
        tooltip: 'Increment',
        child: new Icon(Icons.add),
      ),
    );
  }
}
```

2.5 使用包资源

Flutter 包类似于 Java 语言里的 jar 包，由全球众多开发者共同提供第三方库。例如，网络请求（http）、自定义导航 / 路由处理（fluro）、集成设备 API（如 url_launcher & battery）以及第三方平台 SDK（如 Firebase）等。这使得开发者可以快速构建应用程序，而无须从头造轮子。

1. 包仓库

所有包（package）都会发布到 Dart 的包仓库里，如图 2-3 所示，输入你想使用的包后点击搜索即可。

 提示　包仓库地址为：https://pub.dartlang.org

2. 包使用示例

接下来使用 url_launcher 这个包来详解讲解第三方包的使用，步骤如下。

步骤 1：打开 pubspec.yaml 文件，在 dependencies 下添加包的名称及版本，如图 2-4 箭头指向的内容所示。

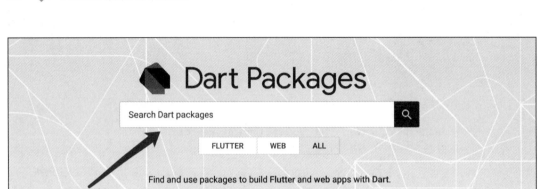

图 2-3　Dart 包仓库

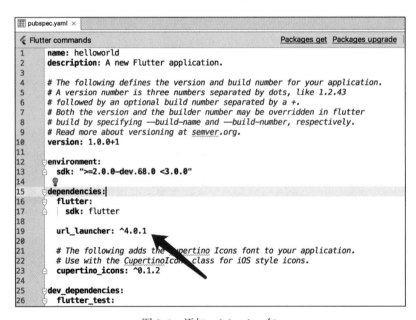

图 2-4　添加 url_lancher 包

步骤 2：点击 Packages get 按钮来获取工程配置文件中所添加的引用包，或者打开命令行窗口执行 flutter packages get 命令，如图 2-5 所示。

 注意　在更新包资源的过程中注意观察控制台消息，可能有版本错误、网络问题，这些都会导致更新失败。

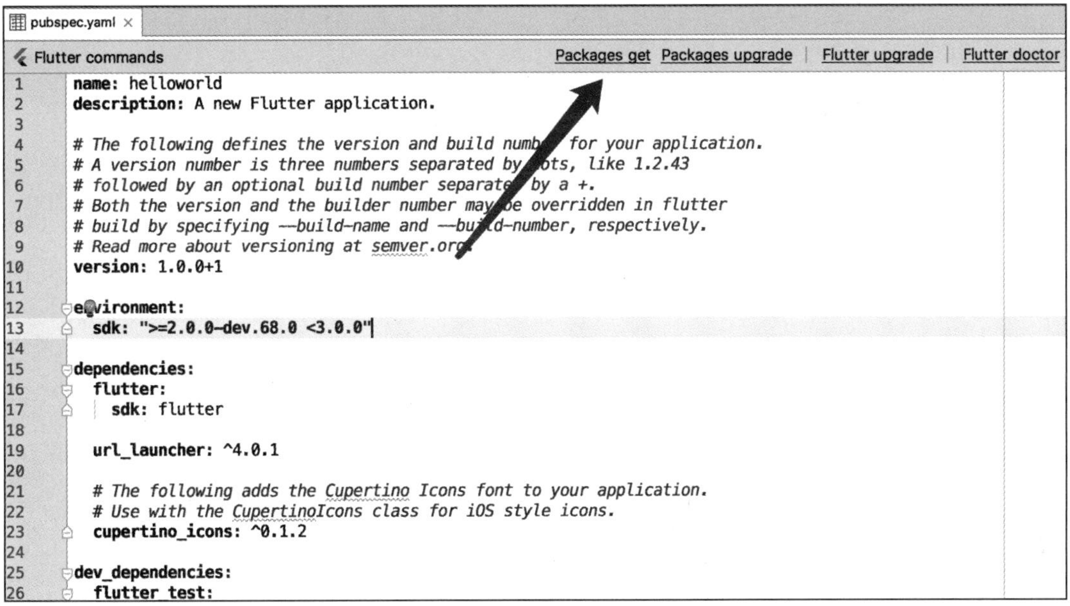

```
pubspec.yaml  ×
Flutter commands                                    Packages get  Packages upgrade  |  Flutter upgrade  |  Flutter doctor
1    name: helloworld
2    description: A new Flutter application.
3
4    # The following defines the version and build number for your application.
5    # A version number is three numbers separated by dots, like 1.2.43
6    # followed by an optional build number separated by a +.
7    # Both the version and the builder number may be overridden in flutter
8    # build by specifying --build-name and --build-number, respectively.
9    # Read more about versioning at semver.org
10   version: 1.0.0+1
11
12   environment:
13     sdk: ">=2.0.0-dev.68.0 <3.0.0"
14
15   dependencies:
16     flutter:
17       sdk: flutter
18
19     url_launcher: ^4.0.1
20
21   # The following adds the Cupertino Icons font to your application.
22   # Use with the CupertinoIcons class for iOS style icons.
23   cupertino_icons: ^0.1.2
24
25   dev_dependencies:
26     flutter_test:
```

图 2-5 执行 Packages get 命令

步骤 3：打开 main.dart 文件，导入 url_launcher.dart 包：

```
import 'package:url_launcher/url_launcher.dart';
```

步骤 4：这时就可以使用 launch 方法来打开 url 地址了：

```
const url = 'https://www.baidu.com';
launch(url);
```

完整的 main.dart 代码如下所示：

```
import 'package:flutter/material.dart';
import 'package:url_launcher/url_launcher.dart';

void main() => runApp(new MyApp());

class MyApp extends StatelessWidget {
  @override
  Widget build(BuildContext context) {

    return new MaterialApp(
      title: '使用第三方包示例',
      home: new Scaffold(
        appBar: new AppBar(
          title: new Text('使用第三方包示例'),
        ),
        body: new Center(
          child: new RaisedButton(
```

```
              onPressed: () {
                // 指定 url 并发起请求
                const url = 'https://www.baidu.com';
                launch(url);
              },
              child: new Text(' 打开百度 '),
            ),
          ),
        ),
      );
    }
  }
```

步骤 5：启动示例，打开界面如图 2-6 所示。

点击"打开百度"按钮，页面会跳转至百度页面，如图 2-7 所示。

图 2-6 使用第三方包示例初始界面　　　　图 2-7 "打开百度"页面效果图

2.6　Http 请求

HTTP 协议通常用于做前后端的数据交互。Flutter 请求网络有两种方法，一种是用 Http 请求，另一种是用 HttpClient 请求。

1. Http 请求方式

在使用 Http 方式请求网络时，需要导入 http 包。如下所示：

```
import 'package:http/http.dart' as http;
```

请看下面的完整示例代码，示例中发起了一个 http 的 get 请求，并将返回的结果信息打印到控制台里：

```
import 'package:flutter/material.dart';
import 'package:http/http.dart' as http;


void main() => runApp(new MyApp());

class MyApp extends StatelessWidget {
  @override
  Widget build(BuildContext context) {
    return new MaterialApp(
      title: 'http 请求示例 ',
      home: new Scaffold(
        appBar: new AppBar(
          title: new Text('http 请求示例 '),
        ),
        body: new Center(
          child: new RaisedButton(
            onPressed: () {

              var url = 'http://httpbin.org/';
              // 向 http://httpbin.org/ 发送 get 请求
              http.get(url).then((response) {
                print(" 状态: ${response.statusCode}");
                print(" 正文: ${response.body}");
              });

            },
            child: new Text(' 发起 http 请求 '),
          ),
        ),
      ),
    );
  }
}
```

请求界面如图 2-8 所示。

点击"发起 http 请求"按钮，程序开始请求指定的 url，如果服务器正常返回数据，则状态码为 200。控制台输出内容如下：

```
Performing hot reload...
Syncing files to device iPhone X...
Reloaded 1 of 509 libraries in 452ms.
```

图 2-8　Http 请求示例效果图

```
flutter: 状态: 200
flutter: 正文: <!DOCTYPE html>
<html lang="en">

<head>
    <meta charset="UTF-8">
    <title>httpbin.org</title>
    <link href="https://fonts.googleapis.com/css?family=Open+Sans:400,700|Source
      +Code+Pro:300,600|Titillium+Web:400,600,700"
        rel="stylesheet">
    <link rel="stylesheet" type="text/css" href="/flasgger_static/swagger-ui.
      css">
    <link rel="icon" type="image/png" href="/static/favicon.ico" sizes="64x64
      32x32 16x16" />
    <style>
        html {
            box-sizing: border-box;
            overflow: -moz-scrollbars-vertical;
            overflow-y: scroll;
        }

        *,
        *:before,
        *:after {
            box-sizing: inherit;
        }

        body {
            margin: 0;
            background: #fafafa;
        }
    </style>
</head>
```

> **注意** 服务器返回状态 200，同时返回正文。完整的正文远不止这些内容，你可以自己测试此示例，并查看控制台的输出消息。

2. HttpClient 请求方式

在使用 HttpClient 方式请求网络时，需要导入 io 及 convert 包，如下所示：

```
import 'dart:convert';
import 'dart:io';
```

请看下面的完整示例代码，示例中使用 HttpClient 请求了一条天气数据，并将返回的结果信息打印到控制台里。具体请求步骤看代码注释即可。

```
import 'package:flutter/material.dart';
import 'dart:convert';
```

```dart
import 'dart:io';

void main() => runApp(MyApp());

class MyApp extends StatelessWidget {
  // 获取天气数据
  void getWeatherData() async {
    try {
      // 实例化一个 HttpClient 对象
      HttpClient httpClient = new HttpClient();

      // 发起请求
      HttpClientRequest request = await httpClient.getUrl(
          Uri.parse("http://t.weather.sojson.com/api/weather/city/101030100"));

      // 等待服务器返回数据
      HttpClientResponse response = await request.close();

      // 使用 utf8.decoder 从 response 里解析数据
      var result = await response.transform(utf8.decoder).join();
      // 输出响应头
      print(result);

      //httpClient 关闭
      httpClient.close();

    } catch (e) {
      print("请求失败: $e");
    } finally {

    }
  }

  @override
  Widget build(BuildContext context) {
    return MaterialApp(
      title: 'httpclient 请求',
      home: Scaffold(
        appBar: AppBar(
          title: Text('httpclient 请求'),
        ),
        body: Center(
          child: RaisedButton(
            child: Text("获取天气数据"),
            onPressed: getWeatherData,
          ),
        ),
      ),
    );
```

                }
            }

请求界面如图 2-9 所示。

图 2-9　HttpClient 请求示例效果图

点击 "获取天气数据" 按钮，程序开始请求指定的 url，如果服务器正常返回数据，则状态码为 200。控制台输出内容如下：

```
Performing hot reload...
Syncing files to device iPhone X...
Reloaded 1 of 419 libraries in 412ms.
flutter: {"time":"2018-11-30 08:09:00","cityInfo":{"city":"天津市","cityId":"101030100","parent":"天津","updateTime":"07:56"},"date":"20181130","message":"Success!", "status":200,"data":{"shidu":"84%","pm25":30.0,"pm10":79.0,"quality":"良","wendu":"1","ganmao":"极少数敏感人群应减少户外活动","yesterday":{"date":"29日星期四","sunrise":"07:08","high":"高温 9.0℃","low":"低温 0.0℃","sunset":"16:50","aqi":86.0,"fx":"东风","fl":"<3级","type":"晴","notice":"愿你拥有比阳光明媚的心情"}, "forecast":[{"date":"30日星期五","sunrise":"07:09","high":"高温 10.0℃","low":"低温 1.0℃","sunset":"16:50","aqi":53.0,"fx":"西风","fl":"<3级","type":"晴","notice":"愿你拥有比阳光明媚的心情"},{"date":"01日星期六","sunrise":"07:10","high":"高温 10.0℃","low":"低温 4.0℃","sunset":"16:49","aqi":94.0,"fx":"东风","fl":"<3级","type":"阴","notice":"不要被阴云遮挡住好心情"},{"date":"02日星期日","sunrise":"07:11","high":"高温 1<…>
```

> **注意** 返回的数据是 JSON 格式,所以后续还需要做 JSON 处理。另外还需要使用 utf8.decoder 从 response 里解析数据。

如果请求里需要带参数,可以在 URI 里增加查询参数,具体的请求地址和参数要根据实际需要编写,代码如下所示:

```
Uri uri=Uri(scheme: "https", host: "t.weather.sojson.com", queryParameters: {
    "_id": 26,
    "city_code": "101030100",// 接口需要的 city_code
    "city_name": " 天津 "
});
```

Chapter 3 第 3 章

Dart 语言简述

在前两章介绍 Flutter 基础知识时,多多少少使用了一些 Dart 语言。作为 Flutter SDK 指定的语言,我们很有必要补充一下 Dart 语言的基础知识,包括它的语法特性、基本语句、面向对象等知识点。

在这一章里将讲解 Dart 语言的知识点:
- Dart 重要概念与常用开发库
- 变量与基本数据类型
- 函数
- 运算符
- 流程控制语句
- 异常处理
- 面向对象
- 泛型
- 库的使用
- 异步支持
- 元数据
- 注释

3.1 Dart 重要概念与常用开发库

Dart 诞生于 2011 年 10 月 10 日,谷歌 Dart 语言项目的领导人 Lars Bak 在丹麦举行的

Goto会议上宣布，Dart是一种"结构化的Web编程"语言，Dart编程语言在所有现代浏览器和环境中提供高性能。

Dart虽然是谷歌开发的计算机编程语言，但后来被ECMA认定为标准。这门语言用于Web、服务器、移动应用和物联网等领域的开发，是宽松开源许可证（修改的BSD证书）下的开源软件。

Dart最新的版本是Dart2，Dart 2是一款高效、简洁、已通过实战检验的语言，能够应对现代应用程序开发的挑战。Dart 2大大加强和精简了类型系统，清理了语法，并重建了大部分开发工具链，使移动和Web开发变得更加愉快和高效。Dart 2还融合了包括Flutter、AdWords和AdSense等工具开发者对该语言早期使用的经验教训，以及针对客户反馈的成千上万大大小小的问题进行了改进。

那么Flutter和Dart有什么关系？确实有关系。早期的Flutter团队评估了十多种语言才选择了Dart，因为它符合构建用户界面的方式。以下是Flutter团队看重Dart语言的部分特性：

- Dart是AOT（Ahead Of Time）编译的，编译成快速、可预测的本地代码，使Flutter几乎都可以使用Dart编写。这不仅使Flutter变得更快，而且几乎所有的组件（包括所有的小部件）都可以定制。
- Dart也可以JIT（Just In Time）编译，开发周期异常快，工作流颠覆常规（包括Flutter流行的亚秒级有状态热重载）。
- Dart可以更轻松地创建以60fps运行的流畅动画和转场。Dart可以在没有锁的情况下进行对象分配和垃圾回收。就像JavaScript一样，Dart避免了抢占式调度和共享内存（因而也不需要锁）。由于Flutter应用程序被编译为本地代码，因此不需要在领域之间建立缓慢的桥梁（例如，JavaScript到本地代码）。它的启动速度也快得多。
- Dart使Flutter不需要单独的声明式布局语言（如JSX或XML），或单独的可视化界面构建器，因为Dart的声明式编程布局易于阅读和可视化。所有的布局使用一种语言，聚集在一处，Flutter很容易提供高级工具，使布局更简单。
- 开发人员发现Dart特别容易学习，因为它具有静态和动态语言用户都熟悉的特性。

并非所有这些功能都是Dart独有的，但Dart将这些功能组合得恰到好处，使Dart在实现Flutter方面独一无二。因此，没有Dart，很难想象Flutter像现在这样强大。

当你想创建移动App、Web App、Command-line应用时，都可以使用Dart语言，如图3-1所示。

Dart重要的概念如下：

- 所有的东西都是对象，无论是变量、数字、函数等都是对象。所有的对象都是类的实例。所有的对象都继承自内置的Object类。这点类似于Java语言"一切皆为对象"。
- 程序中指定数据类型使得程序合理地分配内存空间，并帮助编绎器进行语法检查。但是，指定类型不是必须的。Dart语言是弱数据类型。
- Dart代码在运行前解析。指定数据类型和编译时的常量，可以提高运行速度。

Flutter
Write a mobile app that runs on both iOS and Android.

Web
Write an app that runs in any modern web browser.

Server
Write a command-line app or server-side app.

图 3-1　Dart 支持的平台

- Dart 程序有统一的程序入口：main()。这一点与 Java、C / C++ 语言相像。
- Dart 没有 public、protected 和 private 的概念。私有特性通过变量或函数加上下划线来表示。
- Dart 的工具可以检查出警告信息（warning）和错误信息（errors）。警告信息只是表明代码可能不工作，但是不会妨碍程序运行。错误信息可以是编译时的错误，也可能是运行时的错误。编译时的错误将阻止程序运行，运行时的错误将会以异常（exception）的方式呈现。
- Dart 支持 anync/await 异步处理。
- 关键字（56 个）如下：abstract, do, import, super, as, dynamic, in, switch, assert, else, interface, sync*, enum, implements, is, this, async*, export, library, throw, await, external, mixin, true, break, extends, new, try, case, factory, null, typedef, catch, false, operator, var, class, final, part, void, const, finally, rethrow, while, continue, for, return, with, covariant, get, set, yield*, default, if, static, deferred。

Dart 语言常用库如表 3-1 所示。

表 3-1　Dart 语言常用库

包名	描述
dart:async	异步编程支持，提供 Future 和 Stream 类
dart:collection	对 dart：core 提供更多的集合支持
dart:convert	不同类型（JSON，UTF-8）间的字符编码、解码支持
dart:core	Dart 语言内建的类型、对象以及 dart 语言核心的功能
dart.html	网页开发用到的库
dart.io	文件读写 I/O 相关操作的库
dart:math	数字常量及函数，提供随机数算法
dart:svg	事件和动画的矢量图像支持

其中如下三个开发库的使用频率最高：

- dart:core：核心库，包括 strings、numbers、collections、errors、dates、URIs 等。
- dart:html：网页开发里 DOM 相关的一些库。

❑ dart:io：I/O 命令行使用的 I/O 库。

dart:core 库是 Dart 语言初始已经包含的库，其他的任何库在使用前都需要加上 import 语句。例如，使用 dart:html 可以使用如下的命令：

```
import 'dart:html';
```

使用官方提供的 pub 工具可以安装丰富的第三方库。第三方库的地址为：pub.dartlang.org。

3.2 变量与基本数据类型

在 Dart 里，变量声明使用 var 关键字，如下所示：

```
var name = '小张';
```

在 Dart 语言里一切皆为对象，所以如果没有将变量初始化，那么它的默认值为 null。下面的示例代码判断 name 是否为 null：

```
int name;
if(name == null);
```

1. 常量和固定值

常量及固定值在开发中很常见，比如星期一到星期天、一年 12 个月，这些数据都可以定义成常量形式。

❑ 如果定义的变量不会变化，可以使用 final 或 const 来指明。const 是一个编译时的常量，final 的值只能被设定一次，示例如下：

```
final username = '张三'; // 定义了一个常量
// username = '李四';    // 会引发一个错误
```

第一行代码设置了一个常量，如果第二行进新重新赋值，那么将引发异常。

❑ 通过对 const 类型做四则运算将自动得到一个 const 类型的值。下面的代码会得到一个常量，计算圆的面积：

```
const pi = 3.1415926;
const area = pi * 100 * 100;
```

❑ 可以通过 const 来创建常量的值，就是说 const[] 本身是构造函数，示例代码如下所示：

```
final stars = const [];
const buttons = const [];
```

2. 基本数据类型

Dart 语言常用的基本数据类型包括：Number、String、Boolean、List、Map。

（1）Number 类型

Number 类型包括如下两类：

- int 整形。取值范围：-2^53 到 2^53。
- doble 浮点型。64 位长度的浮点型数据，即双精度浮点型。

int 和 double 类型都是 num 类型的子类。int 类型不能包含小数点。num 类型包括的操作有：+，-，*，/ 以及位移操作 >>。num 类型包括的常用方法有：abs、ceil 和 floor。

（2）String 类型

String 类型也就是字符串类型，在开发中大量使用。定义的例子如下所示：

```
var s1 = 'hello world'; // 单双引号都可以
```

String 类型可以使用 + 操作，非常方便，具体用法示例如下所示：

```
var s1 = 'hi ';

var s2 = 'flutter';

var s3 = s1 + s2;

print(s3);
```

上面代码打印输出"hi flutter"字符串。

可以使用三个单引号或双引号来定义多行的 String 类型，在 Flutter 中我们专门用来表示大文本块。示例代码如下所示：

```
var s1 = '''
请注意这是一个用三个单引号包裹起来的字符串，
可以用来添加多行数据。
''';

var s2 = """同样这也是一个用多行数据，
只不过是用双引号包裹起来的。
""";
```

（3）Boolean 类型

Dart 是强 bool 类型检查，只有 bool 类型的值是 true 才被认为是 true。有的语言里 0 是 false，大于 0 是 true。在 Dart 语言里则不是，值必须为 true 或者 false。下面的示例代码编译不能正常通过，原因是 sex 变量是一个字符串，不能使用条件判断语句，必需使用 bool 类型才可以：

```
var sex = '男';
if (sex) {
  print('你的性别是 !' + sex);
}
```

(4) List 类型

在 Dart 语言中,具有一系列相同类型的数据称为 List 对象。Dart 里的 List 对象类似于 JavaScript 语言的数组 Array 对象。定义 List 的例子如下所示:

```
var list = [1, 2, 3];
```

List 对象的第一个元素的索引是 0,最后一个元素的索引是 list.lenght – 1,代码如下所示:

```
var list = [1,2,3,4,5,6];
print(list.length);
print(list[list.length - 1]);
```

上面的代码输出长度为 6,最后一个元素值也为 6。

(5) Map 类型

Map 类型将 key 和 value 关联在一起,也就是健值对。像其他支持 Map 的编程语言一样,key 必须是唯一的。

如下代码是 Map 对象的定义,示例定义了一个关于星期的键值对对象:

```
var week = {
    'Monday' : '星期一',
    'Tuesday': '星期二',
    'Wednesday' : '星期三',
    'Thursday': '星期四',
    'Friday' : '星期五',
    'Saturday' : '星期六',
    'Sunday' : '星期日',
};
```

也可以使用 Map 对象的构造函数 Map() 来创建 Map 对象,如下所示:

```
var week = new Map();
    week['Monday'] = '星期一';
    week['Tuesday'] = '星期二';
    week['Wednesday'] = '星期三';
    week['Thursday'] = '星期四';
    week['Friday'] = '星期五';
    week['Saturday'] = '星期六';
    week['Sunday'] = '星期日';
```

添加新的 key-value 对,再给 week 添加一个值,注意,其中 0 为键不是数组的下标索引:

```
week['0'] = '星期一';
```

检查 key 是否在 Map 对象中:

```
assert(week['Monday'] == null);
```

使用 length 来获取 key-value 对的数量,现在我们调用 length 输出长度结果为 8,原因是后面又添加了一个数据,代码如下所示:

```
print(week.length);
```

3.3 函数

Dart 是一个面向对象的语言,所以函数也是对象,函数属于 Function 对象。

函数可以像参数一样传递给其他函数,这样便于做回调处理。

如下示例为判断两个字符串是否相等:

```
// 判断两个字符串是否相等
bool equal(String str1, String str2) {
  return str1 == str2;
}
```

1. 可选参数

将参数使用中括号 [] 括起来,用来表明是可选位置参数。例如,总共传入了三个参数,其中 name 和 sex 是必需传入的参数,from 参数可以不传,代码如下:

```
// 获取用户信息
String getUserInfo(String name, String sex, [String from]) {
  var info = '$name 的性别是 $sex';
  if (from != null) {
    info = '$info 来自 $from';
  }
  return info;
}

void test(){
  print(getUserInfo('小王', '男'));
}
```

调用上面的 test 方法可以输出"小王的性别是男",但是不会输出来自哪里。

2. 参数默认值

如果参数指定了默认值,当不传入值时,函数里会使用这个默认值。如果传入了值,则用传入的值取代默认值。通常参数的默认值为 null。改造上面获取用户信息的例子,给 from 参数赋上默认值,具体代码如下:

```
// 获取用户信息 使用等号 (`= `) 来设置默位置字参数
String getUserInfo(String name, String sex, [String from = '中国']) {
  var info = '$name 的性别是 $sex';
  if (from != null) {
    info = '$info 来自 $from';
  }
  return info;
}

void test(){
  print(getUserInfo('小王', '男'));
}
```

调用上面的 test 方法可以输出"小王的性别是男来自中国",这里大家会发现输出了来自哪里,就是因为我们使用了默认参数值。

3. main 函数

Flutter 应用程序必须要有一个 main 函数,和其他语言一样作为程序的入口函数。下面的代码表示应用要启动 MyApp 类:

```
void main() => runApp(MyApp());
```

4. 函数返回值

在 Dart 语言中,函数的返回值有如下特点:
- 所有的函数都会有返回值。
- 如果没有指定函数返回值,则默认的返回值是 null。
- 没有返回值的函数,系统会在最后添加隐式的 return 语句。

3.4 运算符

Dart 支持各种类型的运算符,并且其中的一些操作符还能进行重载。完整的操作符如表 3-2 所示。

使用运算符时可以创建表达式,以下是运算符表达式的一些示例:

```
a++
a--
a + b
a = b
a == b
expr ? a : b
a is T
```

在表 3-2 操作符表中,操作符的优先级由上到下逐个减小,上面行内的操作符优先级大于下面行内的操作符。例如,"乘法类型"操作符 % 的优先级比"等价"操作符 == 要高,而 == 操作符的优先级又比"逻辑与"操作符 && 要高。注意使用运算符时的顺序,方法如下所示:

表 3-2 Dart 的运算符

描述	运算符
一元后缀	expr++ expr-- () [] . ?.
一元前缀	-expr !expr ~expr ++expr --expr
乘法类型	* / % ~/
加法类型	+ -
移动位运算	<< >>
与位运算	&
异或位运算	^
或位运算	|
关系和类型测试	>= <= > < as is is!
等式	== !=
逻辑与	&&
逻辑或	||
条件	expr1 ? expr2 : expr3
级联	..
赋值	= *= /= ~/= %= += -= <<= >>= &= ^= |= ??=

```
// 1.使用括号来提高可读性
if ((n % i == 0) && (d % i == 0))
```

```
// 2. 难以阅读，但是和上面等价
if (n % i == 0 && d % i == 0)
```

 提示　对于二元运算符，其左边的操作数将会决定使用的操作符的种类。例如，当你使用一个 Vector 对象以及一个 Point 对象时，aVector + aPoint 使用的 + 是由 Vector 所定义的。

1. 算术运算符

Dart 支持常用的算术运算符如下所示：

操作符	含义
+	加
–	减
-expr	一元减号，也命名为负号（使后面表达式的值反过来）
*	乘
/	除
~/	返回一个整数值的除法
%	取余，除法剩下的余数

示例代码如下：

```
assert(3 + 6 == 9);
assert(3 - 6 == -3);
assert(3 * 6 == 18);
assert(7 / 2 == 3.5); // 结果是浮点型
assert(5 ~/ 2 == 2); // 结果是整型
assert(5 % 2 == 1); // 求余数
```

Dart 还支持前缀和后缀递增和递减运算符，如下所示：

操作符	含义
++var	var=var+1 表达式的值为 var+1
var++	var=var+1 表达式的值为 var
--var	var=var−1 表达式的值为 var−1
var--	var=var−1 表达式的值为 var

示例代码如下：

```
var a, b;

a = 0;
b = ++a; // 在b获得其值之前自增a
assert(a == b); // 1 == 1

a = 0;
b = a++; // 在b获得值后自增a
assert(a != b); // 1 != 0

a = 0;
b = --a; // 在b获得其值之前自减a
```

```
assert(a == b); // -1 == -1

a = 0;
b = a--; // 在b获得值后自减a
assert(a != b); // -1 != 0
```

2. 关系运算符

等式和关系运算符的含义如下:

操作符	含义
==	等于
!=	不等于
>	大于
<	小于
>=	大于等于
<=	小于等于

有时需要判断两个对象是否相等,请使用==运算符。

下面是使用每个等式和关系运算符的示例:

```
assert(2 == 2);
assert(2 != 3);
assert(3 > 2);
assert(2 < 3);
assert(3 >= 3);
assert(2 <= 3);
```

3. 类型测试操作符

as、is 和 is! 操作符在运行时用于检查类型非常方便,含义如下所示:

操作符	含义
as	类型转换
is	当对象是相应类型时返回 true
is!	当对象不是相应类型时返回 true

如果 obj 实现了 T 所定义的接口,那么 obj is T 将返回 true。

使用 as 操作符可以把一个对象转换为指定类型,前提是能够转换。转换之前用 is 判断一下更保险。如下面这段代码:

```
if (user is User) {
  // 类型检测
  user.name = 'Flutter';
}
```

如果能确定 user 是 User 的实例,则你可以通过 as 直接简化代码:

```
(user as User).name = 'Flutter';
```

> **注意** 上面两段代码并不相等。如果 user 的值为 null 或者不是一个 User 对象,第一段代码不会做任何事情,第二段代码会报错。

4. 赋值操作符

可以使用 = 运算符赋值。要仅在变量为 null 时赋值，请使用 ??= 运算符。如下面代码所示：

```
// 赋值给 a
a = value;
// 如果 b 为空，则将值分配给 b；否则，b 保持不变
b ??= value;
```

诸如 += 之类的复合赋值运算符将操作与赋值相结合。以下是复合赋值运算符的工作方式：

复合赋值	等式表达式
a op b	a = a op b
a += b	a = a + b
a -= b	a = a - b

5. 逻辑运算符

可以使用逻辑运算符反转或组合布尔表达式，逻辑运算符如下所示：

操作符	含义
!expr	反转以下表达式（将 false 更改为 true，反之亦然）
||	逻辑或
&&	逻辑与

示例代码如下：

```
if (!expr && (test == 1 || test == 8)) {
  // ...TODO...
}
```

6. 位运算符

通常我们指位运算为 << 或 >> 移动位运算，通过操作位的移动来达到运算的目的，而 &、|、^、~expr 也是操作位来达到运算的目的。具体含义如下所示：

操作符	含义
&	与
|	或
^	异或
~expr	一元位补码（0s 变为 1s；1s 变为 0s）
<<	左移
>>	右移

示例代码如下：

```
final value = 0x22;
final bitmask = 0x0f;
```

```
assert((value & bitmask)  == 0x02);    // 与
assert((value & ~bitmask) == 0x20);    // 与非
assert((value | bitmask)  == 0x2f);    // 或
assert((value ^ bitmask)  == 0x2d);    // 异或
assert((value << 4)       == 0x220);   // 左移
assert((value >> 4)       == 0x02);    // 右移
```

7. 条件表达式

Dart 有两个运算符，可用来简明地评估可能需要 if-else 语句的表达式。如下代码即为一种条件表达式，也可以称为三元表达式。如果条件为真，返回 expr1，否则返回 expr2：

```
condition ? expr1 : expr2
```

第二种如下所示，如果 expr1 为非空，则返回其值；否则，计算并返回 expr2 的值：

```
expr1 ?? expr2
```

8. 级联操作

级联操作用两个点（..）表示，可对同一对象执行一系列操作。类似于 Java 语言里点点点处理或 JavaScript 里的 Promise 的 then 处理。级联操作主要的目的是为了简化代码，示例如下：

```
querySelector('#btnOK') // 获取一个 id 为 btnOK 的按钮对象
  ..text = ' 确定 ' // 使用它的成员
  ..classes.add('ButtonOKStyle')
  ..onClick.listen((e) => window.alert(' 确定 '));
```

第一个方法调用 querySelector，返回一个按钮对象，然后再设置它的文本为"确定"，再给这个按钮添加一个样式叫 'ButtonOKStyle'，最后再监听单击事件，事件弹出一个显示"确定"的 Alert。这个例子相当于如下操作：

```
var button = querySelector('#btnOK');
button.text = ' 确定 ';
button.classes.add(''ButtonOKStyle'');
button.onClick.listen((e) => window.alert(' 确定 '));
```

> **注意** 严格来说，级联的"双点"符号不是运算符，这只是 Dart 语法的一部分。

3.5 流程控制语句

Dart 语言的流程控制语句如下：

- ❑ if 和 else
- ❑ for（循环）
- ❑ while 和 do-while（循环）

- break 和 continue
- switch 和 case
- assert（断言）
- try-catch 和 throw

1. if 和 else

Dart 支持 if 及 else 的多种组合，示例代码如下：

```
String today = 'Monday';
if (today == 'Monday') {
  print('今天是星期一');
} else if (today == 'Tuesday') {
  print('今天是星期二');
} else {
  print('今天是个好日子');
}
```

上面的代码输出"今天是星期一"，条件语句走到第一条判断就停止了。

2. for

下面举例说明 for 循环，首先定义了一个字符串"Hello Dart"，然后使用 for 循环向 message 变量里写入 5 个同样的字符"!"，如下所示：

```
var message = new StringBuffer("Hello Dart");
for (var i = 0; i < 5; i++) {
  message.write('!');
}
print(message);
```

上面的代码会输出"Hello Dart!!!!!"，注意值是字符串向尾部添加的。除了常规的 for 循环外，针对可以序列化的操作数，可以使用 forEach() 方法。当不关心操作数的当前下标时，forEach() 方法是很简便的。如下代码所示：

```
var arr = [0, 1, 2, 3, 4, 5, 6];
    for (var v in arr) {
      print(v);
    }
```

上面的代码会按序列输出 0 1 2 3 4 5 6。

3. while 和 do-while

下面举例说明 while 循环，其中定义了一个变量 temp，temp 在循环体内自动加 1，当条件（temp ＜ 5）不满足时会退出循环，如下所示：

```
var _temp = 0;
while(_temp < 5){

    print("这是一个while循环：" + (_temp).toString());
```

```
    _temp ++;
}
```

接下来我们看一下例子用 do-while 循环,代码如下所示:

```
var _temp = 0;
    do{
      print(" 这是一个循环: " + (_temp).toString());
      _temp ++;
    }
    while(_temp < 5);
```

上面的两个例子都对应如下输出:

```
flutter: 这是一个循环: 0
flutter: 这是一个循环: 1
flutter: 这是一个循环: 2
flutter: 这是一个循环: 3
flutter: 这是一个循环: 4
```

4. break 和 continue

break 用来跳出循环,改造前面的循环例子,代码如下:

```
var arr = [0, 1, 2, 3, 4, 5, 6];
    for (var v in arr) {
      if(v == 2 ){
        break;
      }
      print(v);
    }
```

上面的代码当 v 等于 2 时循环结束。所以程序输出 0,1。现在我们把 break 改为 continue,代码如下所示:

```
var arr = [0, 1, 2, 3, 4, 5, 6];
    for (var v in arr) {
      if(v == 2 ){
        //break;
        continue;
      }
      print(v);
    }
```

改为 continue 后,当 v 等于 2 时循环只是跳出本次循环,代码还会继续向下执行,所以输出的结果是 0,1,3,4,5,6。

5. switch 和 case

Dart 中 switch / case 语句使用 == 操作来比较整数、字符串或其他编译过程中的常量,从而实现分支的作用。switch / case 语句的前后操作数必须是相同类型的对象实例。每一个非空的 case 子句最后都必须跟上 break 语句。示例如下所示:

```
String today = 'Monday';
    switch (today) {
      case 'Monday':
        print('星期一');
        break;
      case 'Tuesday':
        print('星期二');
        break;
    }
```

上面这段代码也可以用 if /else 语句，输出相同的结果，代码输出为"星期一"。

6. assert

Dart 语言通过使用 assert 语句来中断正常的执行流程，当 assert 判断的条件为 false 时发生中断。assert 判断的条件是任何可以转化为 boolean 类型的对象，即使是函数也可以。如果 assert 的判断为 true，则继续执行下面的语句；反之则会抛出一个断言错误异常 AssertionError。代码如下所示：

```
// 确定变量的值不为 null
assert(text != null);
```

3.6　异常处理

异常是表示发生了意外的错误，如果没有捕获异常，引发异常的隔离程序将被挂起，并且程序将被终止。

Dart 代码可以抛出并捕获异常，但与 Java 相反，Dart 的所有异常都是未检查的异常。方法不声明它们可能抛出哪些异常，也不需要捕获任何异常。

Dart 提供了异常和错误类型以及许多预定义的子类型。当然，也可以定义自己的异常。然而，Dart 程序可以抛出任何非空对象。

1. 抛出异常

下面是一个抛出或引发异常的例子：

```
throw FormatException('抛出一个 FormatException 异常');
```

你也可以抛出任意对象：

```
throw '数据非法！';
```

 稳定健壮的程序一定是做了大量异常处理的，所以建议你在编写程序时尽量考虑到可能发生异常的情况。

2. 捕获异常

你可以指定一个或两个参数来捕获异常（catch），第一个是抛出的异常，第二个是堆栈跟踪（StackTrace 对象）。如下面代码所示：

```
try {
    // ...
} on Exception catch (e) {
    print('Exception details:\n $e');
} catch (e, s) {
    print('Exception details:\n $e');
    print('Stack trace:\n $s');
}
```

上面的代码第一个 catch 用来捕获异常详细信息，第二个 catch 是堆栈跟踪信息。

3. Finally

要确保某些代码能够运行，无论是否抛出异常，请使用 finally 子句。如果没有 catch 子句匹配异常，则异常在 finally 子句运行后传播。如下面代码所示，在最下加上了 finally 语句：

```
try {
    // ...
} on Exception catch (e) {
    print('Exception details:\n $e');
} catch (e, s) {
    print('Exception details:\n $e');
    print('Stack trace:\n $s');
} finally {
    print('Do some thing:\n');
}
```

3.7 面向对象

Dart 作为高级语言支持面向对象的很多特性，并且支持基于 mixin 的继承方式。基于 mixin 的继承方式是指：一个类可以继承自多个父类，相当于其他语言里的多继承。所有的类都有同一个基类 Object，这个特性类似于 Java 语言，Java 所有的类也都是继承自 Object，也就是说一切皆为对象。

使用 new 语句实例化一个类，如下所示：

```
// 实例化了一个 User 类的对象 user
var user = new User('张三', 20);
```

3.7.1 实例化成员变量

定义一个 User 类，在里类里添加两个成员变量 name 与 age，代码如下所示：

```
class User{
    String name; //name 成员变量
```

```
    int age; //age 成员变量
}
```

类定义中所有的变量都会隐式的定义 setter 方法，针对非空的变量会额外增加 getter 方法。实例化成员变量请参考如下代码：

```
class User{
  String name; //name 成员变量
  int age; //age 成员变量
}

main() {
  var user = new User();
  user.name = '张三';// 相当于使用了 name 的 setter 方法
  user.age = 20;

}
```

3.7.2 构造函数

1. 常规构造函数

构造函数是用来构造当前类的函数，是一种特殊的函数，函数名称必须要和类名相同才行。如下代码为 User 类添加了一个构造函数，函数里给 User 类的两个成员变量初始化了值：

```
class User{

  String name;
  int age;

  User(String name,int age){
    this.name = name;
    this.age = age;
  }

}
```

this 关键字指向了当前类的实例。上面的代码可以简化为：

```
class User{

  String name;
  int age;

  User(this.name,this.age);

}
```

2. 命名的构造函数

使用命名构造函数从另一类或现有的数据中快速实现构造函数，代码如下所示：

```
class User {
  String name;
  int age;

  User(this.name, this.age);

  // 命名的构造函数
  User.fromJson(Map json) {
    name = json['name'];
    age = json['age'];
  }
}
```

3. 构造函数初始化列表

除了调用父类的构造函数，也可以通过初始化列表在子类的构造函数运行前来初始化实例的成员变量值，代码如下所示：

```
class User {
  final String name;
  final int age;

  User(name, age)
      : name = name,
        age = age;
}

main() {
  var p = new User('张三', 20);
}
```

3.7.3　读取和写入对象

get() 和 set() 方法是专门用于读取和写入对象的属性的方法，每一个类的实例，系统都隐式地包含有 get() 和 set() 方法。这和很多语言里的 VO 类相似。

例如，定义一个矩形的类，有上、下、左、右四个成员变量：top、bottom、left、right，使用 get 及 set 关键字分别对 right 及 bottom 进行获取和设置值。代码如下所示：

```
class Rectangle {
  num left;
  num top;
  num width;
  num height;

  Rectangle(this.left, this.top, this.width, this.height);

  // 获取right值
  num get right             => left + width;

  // 设置right值 同时left也发生变化
```

```
    set right(num value)  => left = value - width;

    // 获取 bottom 值
    num get bottom         => top + height;

    // 设置 bottom 值 同时 top 也发生变化
    set bottom(num value) => top = value - height;
}

main() {
  var rect = new Rectangle(3, 4, 20, 15);

  print('left:'+rect.left.toString());
  print('right:'+rect.right.toString());
  rect.right = 30;
  print('更改 right 值为 30');
  print('left:'+rect.left.toString());
  print('right:'+rect.right.toString());

  print('top:'+rect.top.toString());
  print('bottom:'+rect.bottom.toString());
  rect.bottom = 50;
  print('更改 bottom 值为 50');
  print('top:'+rect.top.toString());
  print('bottom:'+rect.bottom.toString());
}
```

上面例子对应的输出为：

```
flutter: left:3
flutter: right:23
flutter: 更改 right 值为 30
flutter: left:10
flutter: right:30
flutter: top:4
flutter: bottom:19
flutter: 更改 bottom 值为 50
flutter: top:35
flutter: bottom:50
```

3.7.4 重载操作

编写一个例子，定义一个 Vector 向量类，编写两个方法分别用于重载加号及减号，那么当两个向量相加，就表示它们的 x 值及 y 值相加，当两个向量相减，就表示它们的 x 值及 y 值相减。完整的示例代码如下：

```
// 定义一个向量类
class Vector {
  final int x;
  final int y;
```

```
    const Vector(this.x, this.y);

    // 重载加号 + (a + b).
    Vector operator +(Vector v) {
      return new Vector(x + v.x, y + v.y);
    }

    // 重载减号 - (a - b).
    Vector operator -(Vector v) {
      return new Vector(x - v.x, y - v.y);
    }
}

main() {
    // 实例化两个向量
    final v = new Vector(2, 3);
    final w = new Vector(2, 2);

    final r1 = v + w;
    print('r1.x='+r1.x.toString() + ' r1.y=' + r1.y.toString());

    final r2 = v - w;
    print('r2.x='+r2.x.toString() + ' r2.y=' + r2.y.toString());
}
```

上面代码的输出结果为：

```
flutter: r1.x=4 r1.y=5
flutter: r2.x=0 r2.y=1
```

3.7.5 继承类

继承是面向对象编程技术的一块基石，因为它允许创建分等级层次的类。继承就是子类继承父类的特征和行为，使得子类对象（实例）具有父类的实例域和方法；或子类从父类继承方法，使得子类具有父类相同的行为。Dart 里使用 extends 关键字来创建一个子类，super 关键子来指定父类。

接下来定义一个动物类，动物具有吃和跑两种能力。再定义一个人类，人类是属于动物类的，人类不仅会吃和会跑，人类还会说、会学习。所以人类相当于动物类的一个扩展。完整的示例如下所示：

```
// 动物类
class Animal {

    // 动物会吃
    void eat(){
        print(' 动物会吃 ');
    }
```

```
    // 动物会跑
    void run(){
      print('动物会跑');
    }
}
// 人类
class Human extends Animal {

    // 人类会说
    void say(){
      print('人类会说');
    }

    // 人类会学习
    void study(){
      print('人类会学习');
    }
}


main() {

    print('实例化一个动物类');
    var animal = new Animal();
    animal.eat();
    animal.run();

    print('实例化一个人类');
    var human = new Human();
    human.eat();
    human.run();
    human.say();
    human.study();
}
```

上面的例子输出结果如下：

```
flutter: 实例化一个动物类
flutter: 动物会吃
flutter: 动物会跑
flutter: 实例化一个人类
flutter: 动物会吃
flutter: 动物会跑
flutter: 人类会说
flutter: 人类会学习
```

3.7.6 抽象类

抽象类类似于 Java 语言中的接口。抽象类里不具体实现方法，只是写好定义接口，具体实现留着调用的人去实现。抽象类可以使用 abstract 关键字定义类。

接下来写一个数据库操作的抽象类的例子。定义一个抽象类叫 DataBaseOperate，里面定义 4 个数据库常用的操作方法"增删改查"。再定义一个类命名为 DataBaseOperateImpl 继承自 DataBaseOperate 用来实现抽象类里的方法。完整的代码如下所示：

```dart
// 数据库操作抽象类
abstract class DataBaseOperate {
  void insert(); // 定义插入的方法
  void delete(); // 定义删除的方法
  void update(); // 定义更新的方法
  void query(); // 定义一个查询的方法
}

// 数据库操作实现类
class DataBaseOperateImpl extends DateBaseOperate {

  // 实现了插入的方法
  void insert(){
    print('实现了插入的方法');
  }

  // 实现了删除的方法
  void delete(){
    print('实现了删除的方法');
  }

  // 实现了更新的方法
  void update(){
    print('实现了更新的方法');
  }

  // 实现了一个查询的方法
  void query(){
    print('实现了一个查询的方法');
  }

}


main() {

  var db = new DataBaseOperateImpl();
  db.insert();
  db.delete();
  db.update();
  db.query();

}
```

上述代码输出结果为：

```
flutter: 实现了插入的方法
flutter: 实现了删除的方法
flutter: 实现了更新的方法
flutter: 实现了一个查询的方法
```

3.7.7 枚举类型

枚举类型是一种特殊的类，通常用来表示相同类型的一组常量值。每个枚举类型都用于一个 index 的 getter，用来标记元素的元素位置。第一个枚举元素的索引是 0：

```
enum Color {
  red,
  green,
  blue
}
```

获取枚举类中所有的值，使用 value 常数：

```
List<Color> colors = Color.values;
```

因为枚举类里面的每个元素都是相同类型，可以使用 switch 语句来针对不同的值做不同的处理，示例代码如下：

```
enum Color {
  red,
  green,
  blue
}
// 定义一个颜色变量 默认值为蓝色
Color aColor = Color.blue;
switch (aColor) {
  case Color.red:
    print('红色');
    break;
  case Color.green:
    print('绿色');
    break;
  default: // 默认颜色
    print(aColor);    // 'Color.blue'
}
```

3.7.8 Mixins

Mixins（混入功能）相当于多继承，也就是说可以继承多个类。使用 with 关键字来实现 Mixins 的功能，示例代码如下所示：

```
class S {
  a() {print("S.a");}
}
```

```
class A {
  a(){print("A.a");}
  b(){print("A.b");}
}

class T = A with S;

main(List<String> args) {
  T t = new T();
  t.a();
  t.b();
}
```

上面代码的输出内容如下所示,从结果上来看 T 具有了 S 及 A 两个类的方法:

```
S.a
A.b
```

3.8 泛型

泛型通常是为了类型安全而设计的,适当地指定泛型类型会生成更好的代码,可以使用泛型来减少代码重复。Dart 中使用 <T> 的方式来定义泛型。例如,如果想要 List 只包含字符串,可以将其声明为 list <String>。如下所示:

```
var names = new List<String>();
names.addAll(['张三','李四','王五']);
```

1. 用于集合类型

泛型用于 List 和 Map 类型参数化:

```
List: <type>
Map: <keyType, valueType>
```

例子代码如下:

```
var names = <String>['张三','李四','王五'];
var weeks = <String, String>{
    'Monday' : '星期一',
    'Tuesday': '星期二',
    'Wednesday' : '星期三',
    'Thursday': '星期四',
    'Friday': '星期五',
    'Saturday': '星期六',
    'Sunday': '星期日',
};
```

2. 在构造函数中参数化

Map 类型的例子如下:

```
var users = new Map<String, User>();
```

3.9 库的使用

1. 引用库

通过 import 语句在一个库中引用另一个库的文件。需要注意以下事项：

- 在 import 语句后面需要接上库文件的路径。
- 对 Dart 语言提供的库文件使用 dart:xx 格式。
- 第三方的库文件使用 package:xx 格式。

import 的例子如下：

```
import 'dart:io';
import 'package:mylib/mylib.dart';
import 'package:utils/utils.dart';
```

2. 指定一个库的前缀

当引用的库拥有相互冲突的名字，可以为其中一个或几个指定不一样的前缀。这与命名空间的概念比较接近，示例代码如下：

```
import 'package:lib1/lib1.dart';
import 'package:lib2/lib2.dart' as lib2;
// ...
Element element1 = new Element();              // 使用 lib1 中的 Element
lib2.Element element2 = new lib2.Element();    // 使用 lib2 中的 Element
```

lib1/lib1.dart 及 lib2/lib2.dart 里都有 Element 类，如果直接引用就不知道具体引用哪个 Element 类，所以代码中把 lib2/lib2.dart 指定成 lib2，这样使用 lib2.Element 就不会发生冲突。

3. 引用库的一部分

如果只需要使用库的一部分内容，可以有选择地引用，有如下关键字：

- show 关键字：只引用一点。
- hide 关键字：除此之外都引用。

示例代码如下：

```
// 导入 foo
import 'package:lib1/lib1.dart' show foo;

// 除了 foo 导入其他所有内容
import 'package:lib2/lib2.dart' hide foo;
```

代码中的第一行只引用 lib1.dart 下的 foo 部分，第二行代码引用 lib2.dart 下的所有内容除了 foo。

3.10 异步支持

Dart 语言是目前少数几个支持异步操作的语言。一般使用 async 函数和 await 表达式实现异步操作。

Dart 库提供 asynchronous 功能，该功能提供接口来进行耗费时间的操作，比如文件读写、网络请求。该功能返回 Future 或 Stream 对象。

可以通过如下的方式来获取 asynchronous 功能返回的 Future 对象的值：

- 使用 async 函数和 await 表达式。
- 使用 Future 功能提供的 API。

可以通过如下的方式来获取 asynchronous 功能返回的 Stream 对象的值：

- 使用 async 和一个异步的循环 (await for)。
- 使用 Stream 的相关 API。

下面的示例代码使用了 async 或 await 异步处理，虽然代码看起来像是同步处理的：

```
await readFile()
```

必须在一个使用了 async 关键字标记后的函数中使用 await 表达式：

```
fileOperate () async {
  // 读取文件
  var file = await readFile();
  // 其他处理
}
```

3.11 元数据

使用元数据给代码添加更多的信息。元数据是以 @ 开始的修饰符，在 @ 后面接着编译时的常量或调用一个常量构造函数。目前 Dart 语言提供三个 @ 修饰符：

- @deprecated 被弃用的。
- @override 重写。
- @proxy 代理。

使用 @override 修饰符可以重写父类方法。改造之前写的例子，人类重写 eat 方法，代码如下所示：

```
// 动物类
class Animal {

  // 动物会吃
  void eat(){
    print('动物会吃');
  }

  // 动物会跑
```

```
  void run(){
    print('动物会跑');
  }
}
// 人类
class Human extends Animal {

  // 人类会说
  void say(){
    print('人类会说');
  }

  // 人类会学习
  void study(){
    print('人类会学习');
  }

  @override
  // 人类也会吃
  void eat(){
    print('人类也会吃');
  }
}


main() {

  print('实例化一个动物类');
  var animal = new Animal();
  animal.eat();
  animal.run();

  print('实例化一个人类');
  var human = new Human();
  human.eat();
  human.run();
  human.say();
  human.study();
}
```

上面的代码输出结果如下，会输出一句"人类也会吃"，表明重写了父类的方法：

```
flutter: 实例化一个动物类
flutter: 动物会吃
flutter: 动物会跑
flutter: 实例化一个人类
flutter: 人类也会吃
flutter: 动物会跑
flutter: 人类会说
flutter: 人类会学习
```

元数据可以修饰 library（库）、class（类）、typedef（类型定义）、type parameter（类型参数）、constructor（构造函数）、factory（工厂函数）、function（函数）、field（作用域）、parameter（参数）、variable declaration（变量声明）。

3.12 注释

Dart 支持三种注释类型：单行注释、多行注释、文档注释。

1. 单行注释

单行注释以 // 开头，从 // 开始到一行结束的所有内容都会被 Dart 编译器忽略，示例代码如下：

```
main() {
  // 打印输出
  print('Hi Dart);
}
```

2. 多行注释

单行注释以 /* 开头，以 */ 结束，之间的所有内容都会被 Dart 编译器忽略掉，示例代码如下：

```
/**
 * 多行注释
 * print(' 实例化一个动物类 ');
    var animal = new Animal();
    animal.eat();
    animal.run();

    print(' 实例化一个人类 ');
    var human = new Human();
    human.eat();
    human.run();
    human.say();
    human.study();
 *
 */
```

3. 文档注释

文档注释以 /** 或 /// 开头，示例代码如下：

```
///print(' 实例化一个动物类 ');
var animal = new Animal();
animal.eat();
animal.run();
/**
 *
 * print(' 实例化一个人类 ');
    var human = new Human();
    human.eat();
    human.run();
    human.say();
    human.study();
 *
 */
```

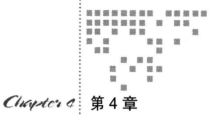

Chapter 4 第 4 章

常 用 组 件

Flutter 里有一个非常重要的核心理念：一切皆为组件，Flutter 所有的元素皆由组件组成。比如：一个布局元素、一个动画、一个装饰效果等。本章主要讲解开发中用得最频繁得组件。

本章所涉及的组件有：
- 容器组件
- 图片组件
- 文本组件
- 图标组件
- 图标按钮组件
- 凸起按钮组件
- 基础列表组件
- 水平列表组件
- 长列表组件
- 网格列表组件
- 表单组件

4.1 容器组件

容器组件（Container）包含一个子 Widget，自身具备如 alignment、padding 等基础属性，方便布局过程中摆放 child。Container 组件常用属性见表 4-1。

表 4-1　Container 组件的属性及描述

属性名	类型	说明
key	Key	Container 唯一标识符，用于查找更新
alignment	AlignmentGeometry	控制 child 的对齐方式，如果 Container 或者 Container 父节点尺寸大于 child 的尺寸，这个属性设置会起作用，有很多种对齐方式
padding	EdgeInsetsGeometry	Decoration 内部的空白区域，如果有 child 的话，child 位于 padding 内部
color	Color	用来设置 Container 背景色，如果 foregroundDecoration 设置的话，可能会遮盖 color 效果
decoration	Decoration	绘制在 child 后面的装饰，设置了 Decoration 的话，就不能设置 color 属性，否则会报错，此时应该在 Decoration 中进行颜色的设置
foregroundDecoration	Decoration	绘制在 child 前面的装饰
width	double	Container 的宽度，设置为 double.infinity 可以强制在宽度上撑满，不设置，则根据 child 和父节点两者一起布局
height	double	Container 的高度，设置为 double.infinity 可以强制在高度上撑满
constraints	BoxConstraints	添加到 child 上额外的约束条件
margin	EdgeInsetsGeometry	围绕在 Decoration 和 child 之外的空白区域，不属于内容区域
transform	Matrix4	设置 Container 的变换矩阵，类型为 Matrix4
child	Widget	Container 中的内容 Widget

 提示　padding 与 margin 的不同之处在于，padding 是包含在 Content 内，而 margin 则是外部边界。设置点击事件的话，padding 区域会响应，而 margin 区域不会响应。

接下来我们编写一个带有装饰效果的 Container 容器组件示例，主要是加了一个边框及底色，示例代码如下：

```
import 'package:flutter/material.dart';

void main() => runApp(MyApp());

class MyApp extends StatelessWidget {
  @override
  Widget build(BuildContext context) {
    return MaterialApp(
      title: '容器组件示例',
      home: Scaffold(
        appBar: AppBar(
          title: Text('容器组件示例'),
        ),
        body: Center(
          // 添加容器
          child: Container(
```

```
          width: 200.0,
          height: 200.0,
          // 添加边框装饰效果
          decoration: BoxDecoration(
            color: Colors.white,
            // 设置上下左右四个边框样式
            border: new Border.all(
              color: Colors.grey, // 边框颜色
              width: 8.0, // 边框粗细
            ),
            borderRadius:
                const BorderRadius.all(const Radius.circular(8.0)), // 边框的弧度
          ),
          child: Text(
            'Flutter',
            textAlign: TextAlign.center,
            style: TextStyle(fontSize: 28.0),
          ),
        ),
      ),
    ),
  );
 }
}
```

上述示例代码视图展现大致如图 4-1 所示。

4.2 图片组件

图片组件（Image）是显示图像的组件，Image 组件有多种构造函数：

❏ new Image：从 ImageProvider 获取图像。

❏ new Image.asset：加载资源图片。

❏ new Image.file：加载本地图片文件。

❏ new Image.network：加载网络图片。

❏ new Image.memory：加载 Uint8List 资源图片。

Image 组件常见属性见表 4-2。

图 4-1 Container 组件应用示例

表 4-2 Image 组件属性及描述

属性名	类型	说明
image	ImageProvider	抽象类，需要自己实现获取图片数据的操作
width/height	double	Image 显示区域的宽度和高度设置，这里需要把 Image 和图片两个东西区分开，图片本身有大小，Image Widget 是图片的容器，本身也有大小。宽度和高度的配置经常和 fit 属性配合使用

(续)

属性名	类型	说明
fit	BoxFit	图片填充模式，具体取值见下一个表
color	Color	图片颜色
colorBlendMode	BlendMode	在对图片进行手动处理的时候，可能用到图层混合，如改变图片的颜色。此属性可以对颜色进行混合处理。有多种模式
alignment	Alignment	控制图片的摆放位置，比如图片放置在右下角则为 Alignment.bottomRight
repeat	ImageRepeat	此属性可以设置图片重复模式。noRepeat 为不重复，Repeat 为 x 和 y 方向重复，repeatX 为 x 方向重复，repeatY 为 y 方向重复
centerSlice	Rect	当图片需要被拉伸显示时，centerSlice 定义的矩形区域会被拉伸，可以理解成我们在图片内部定义一个点 9 个点文件用作拉伸，9 个点为"上""下""左""右""上中""下中""左中""右中""正中"
matchTextDirection	bool	matchTextDirection 与 Directionality 配合使用。TextDirection 有两个值分别为：TextDirection.ltr 从左向右展示图片，TextDirection.rtl 从右向左展示图片
gaplessPlayback	bool	当 ImageProvider 发生变化后，重新加载图片的过程中，原图片的展示是否保留。值为 true 则保留；值为 false 则不保留，直接空白等待下一张图片加载

BoxFit 取值及描述参见表 4-3。

表 4-3 BoxFit 取值及描述

取值	描述
BoxFit.fill	全图显示，显示可能拉伸，充满
BoxFit.contain	全图显示，显示原比例，不需充满
BoxFit.cover	显示可能拉伸，可能裁剪，充满
BoxFit.fitWidth	显示可能拉伸，可能裁剪，宽度充满
BoxFit.fitHeight	显示可能拉伸，可能裁剪，高度充满
BoxFit.none	原始大小
BoxFit.scaleDown	效果和 BoxFit.contain 差不多，但是此属性不允许显示超过源图片大小，即可小不可大

下面的示例加载了一张网络图片，以 BoxFit.fitWidth 模式进行填充图片。

示例代码如下：

```
import 'package:flutter/material.dart';

void main() {
  runApp(
    new MaterialApp(
      title: 'Image demo',
      home: new ImageDemo(),
    )
  );
}
```

```
class ImageDemo extends StatelessWidget {
  @override
  Widget build(BuildContext context) {
    return new Center(
      // 添加网络图片
      child: new Image.network(
        // 图片 url
        'https://flutter.io/images/flutter-mark-
         square-100.png',
        // 填充模式
        fit: BoxFit.fitWidth,
      ),
    );
  }
}
```

上述示例代码视图展现大致如图 4-2 所示。

4.3 文本组件

文本组件（text）负责显示文本和定义显示样式，常用属性见表 4-4。

接下来我们通过创建多个文本组件来展示不同的文本样式。比如不同的颜色、不同的字号、不同的线形等，完整示例代码如下：

图 4-2 Image 组件应用示例

表 4-4 Text 组件属性及描述

属性名	类型	默认值	说明
data	String		数据为要显示的文本
maxLines	int	0	文本显示的最大行数
style	TextStyle	null	文本样式，可定义文本的字体大小、颜色、粗细等
textAlign	TextAlign	TextAlign.center	文本水平方向对齐方式，取值有 center、end、justify、left、right、start、values
textDirection	TextDirection	TextDirection.ltr	有些文本书写的方向是从左到右，比如英语、泰米尔语、中文，有些则是从右到左，比如亚拉姆语、希伯来语、乌尔都语。从左到右使用 TextDirection.ltr，从右到左使用 TextDirection.rtl
textScaleFactor	double	1.0	字体缩放系数，比如，如果此属性设置的值为 1.5，那么字体会放大到 150%，也就是说比原来的大了 50%
textSpan	TextSpan	null	文本块，TextSpan 里可以包含文本内容及样式

```
import 'package:flutter/material.dart';

class ContainerDemo extends StatelessWidget {
  @override
  Widget build(BuildContext context) {
```

```
return new Scaffold(
  appBar: new AppBar(
    title: new Text(' 文本组件 '),
  ),
  body: new Column(
    children: <Widget>[

      new Text(
        ' 红色 + 黑色删除线 +25 号 ',
        style: new TextStyle(
          color: const Color(0xffff0000),
          decoration: TextDecoration.lineThrough,
          decorationColor: const Color(0xff000000),
          fontSize: 25.0,

        ),
      ),
      new Text(
        ' 橙色 + 下划线 +24 号 ',
        style: new TextStyle(
          color: const Color(0xffff9900),
          decoration: TextDecoration.underline,
          fontSize: 24.0,

        ),
      ),

      new Text(
        ' 虚线上划线 +23 号 + 倾斜 ',
        style: new TextStyle(
          decoration: TextDecoration.overline,
          decorationStyle: TextDecorationStyle.dashed,
          fontSize: 23.0,
          fontStyle: FontStyle.italic,
        ),
      ),
      new Text(
        '24 号 + 加粗 ',
        style: new TextStyle(
          fontSize: 23.0,
          fontStyle: FontStyle.italic,
          fontWeight: FontWeight.bold,
          letterSpacing: 6.0,
        ),
      ),

    ],
  ),
```

```
      );
    }
  }
  void main() {
    runApp(
      new MaterialApp(
        title: 'Text demo',
        home: new ContainerDemo(),
      )
    );
  }
```

上述示例代码视图展现大致如图 4-3 所示。

4.4 图标及按钮组件

4.4.1 图标组件

图标组件（Icon）为展示图标的组件，该组件不可交互，要实现可交互的图标，可以考虑使用 IconButton 组件。图标组件相关的组件有以下几个：

- IconButton：可交互的 Icon。
- Icons：框架自带 Icon 集合。
- IconTheme：Icon 主题。
- ImageIcon：通过 AssetImages 或者其他图片显示 Icon。

图 4-3 Text 组件应用示例

图标组件常用属性见表 4-5。

表 4-5 Icon 组件属性及描述

属性名	类型	默认值	说明
color	Color	null	图标的颜色，例如 Colors.green[500]
icon	IconData	null	展示的具体图标，可以使用 Icons 图标列表中的任意一个图标即可，如 Icons.phone 表示一个电话的图标
style	TextStyle	null	文本样式，可定义文本的字体大小、颜色、粗细等
size	Double	24.0	图标的大小，注意需要带上小数位
textDirection	TextDirection	TextDirection.ltr	Icon 组件里也可以添加文本内容。有些文本书写的方向是从左到右，有些则是从右到左。从左到右使用 TextDirection.ltr，从右到左使用 TextDirection.rtl

接下来通过一个示例来展示图标组件的使用，Icon 实例化需要传入几个主要的参数，分别为图标、颜色、大小。其中 Icon 的实例化代码如下：

```
new Icon(Icons.phone,color: Colors.green[500],size: 80.0,)
```

完整的示例代码如下：

```
import 'package:flutter/material.dart';

void main() => runApp(
  new MaterialApp(
    title: '图标组件示例',
    home: new LayoutDemo(),
  ),
);

class LayoutDemo extends StatelessWidget {

  @override
  Widget build(BuildContext context) {

    return new Scaffold(
      appBar: new AppBar(
        title: new Text('图标组件示例'),
      ),
      body: new Icon(Icons.phone,color: Colors.
        green[500],size: 80.0,),
    );

  }
}
```

上述示例代码视图展现大致如图 4-4 所示。

图 4-4　图标组件应用示例

4.4.2　图标按钮组件

图标按钮组件（IconButton）是基于 Material Design 风格的组件，它可以响应按下事件，并且按下时会带一个水波纹的效果。如果它的 onPressed 回调函数为 null，那么这个按钮处于禁用状态，并且不可以按下。常用属性见表 4-6。

表 4-6　IconButton 组件属性及描述

属性名	类型	默认值	说明
alignment	AlignmentGeometry	Alignment.center	定义 IconButton 的 Icon 对齐方式，默认为居中。Alignment 是可以设置 x，y 的偏移量的
icon	Widget	null	展示的具体图标，可以使用 Icons 图标列表中的任意一个图标即可，如 Icons.phone 表示一个电话的图标
color	Color	null	图标组件的颜色

（续）

属性名	类型	默认值	说明
disabledColor	Color	ThemeData.disabledColor	图标组件禁用的颜色，默认为主题里的禁用颜色，也可以设置为其他颜色值
iconSize	double	24.0	图标的大小，注意需要带上小数位
onPressed	VoidCallback	null	当按钮按下时会触发此回调事件
tooltip	String	" "	当按钮按下时的组件提示语句

示例代码如下：

```dart
import 'package:flutter/material.dart';

void main() => runApp(
  new MaterialApp(
    title: '图标按钮组件示例',
    home: new LayoutDemo(),
  ),
);

class LayoutDemo extends StatelessWidget {

  @override
  Widget build(BuildContext context) {

    return new Scaffold(
      appBar: new AppBar(
        title: new Text('图标按钮组件示例'),
      ),
      body: new Center(
        child: new IconButton(
          icon: Icon(Icons.volume_up,size: 48.0,),
          tooltip: '按下操作',
          onPressed: () {
            print('按下操作');
          },
        ),
      ),
    );

  }
}
```

上述示例代码视图展现大致如图4-5所示。

当按下图中的喇叭小按钮后，会触发onPressed事件，控制台打印输出内容，如图4-6所示。

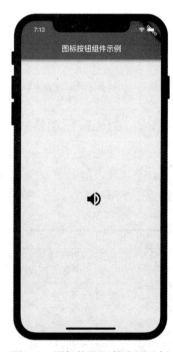

图4-5 图标按钮组件应用示例

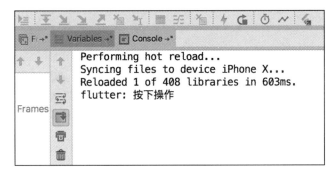

图 4-6 图标按钮组件控制台输出图

4.4.3 凸起按钮组件

凸起按钮组件（RaisedButton）是 Material Design 中的 button，一个凸起的材质矩形按钮，它可以响应按下事件，并且按下时会带一个触摸的效果。

RaisedButton 组件常用属性见表 4-7。

表 4-7 RaisedButton 组件属性及描述

属性名	类型	默认值	说明
color	Color	null	组件的颜色
disabledColor	Color	ThemeData.disabledColor	组件禁用的颜色，默认为主题里的禁用颜色，也可以设置为其他颜色值
onPressed	VoidCallback	null	当按钮按下时会触发此回调事件
child	Widget		按钮的 child 通常为一个 Text 文本组件，用来显示按钮的文本
enable	bool	true	按钮是否为禁用状态

示例代码如下：

```
import 'package:flutter/material.dart';

void main() => runApp(new MyApp());

class MyApp extends StatelessWidget {
  @override
  Widget build(BuildContext context) {
    return new MaterialApp(
      title: 'RaisedButton 示例 ',
      home: new Scaffold(
        appBar: new AppBar(
          title: new Text('RaisedButton 组件示例 '),
        ),
        body: new Center(
          child: new RaisedButton(
            onPressed: () {
```

```
        // 按下事件处理
      },
      child: new Text('RaisedButton组件'),
    ),
   ),
  ),
 );
 }
}
```

上述示例代码视图展现大致如图4-7所示,当按下图中按钮后,会触发onPressed事件,这和IconButton一样。

4.5 列表组件

列表是前端是最常见的需求。在Flutter中,用ListView来显示列表项,支持垂直和水平方向展示,通过一个属性我们就可以控制其方向,列表有以下分类:

❑ 水平的列表
❑ 垂直的列表
❑ 数据量非常大的列表
❑ 矩阵式的列表

图4-7 RaisedButton组件应用示例

4.5.1 基础列表组件

基础列表组件为ListView组件,其常用属性见表4-8。

表4-8 ListView组件属性及描述

属性名	类型	默认值	说明
scrollDirection	Axis	Axis.vertical	列表的排列方向,Axis.vertical为垂直方向,是默认值,Axis.horizontal为水平方向
padding	EdgeInsetsGeometry		列表内部的空白区域,如果有child的话,child位于padding内部
reverse	bool	false	组件排列反向
children	List<Widget>		列表元素,注意List元素全部为Widget类型

示例代码如下:

```
import 'package:flutter/material.dart';


void main() => runApp(new MyApp());
```

```
class MyApp extends StatelessWidget {
  @override
  Widget build(BuildContext context) {

    final title = "基础列表示例";

    return MaterialApp(
      title: title,
      home: Scaffold(
        appBar: AppBar(
          title: Text(title),
        ),
        //添加基础列表
        body: new ListView(
          //添加静态数据
          children: <Widget>[
            ListTile(
              //添加图标
              leading: Icon(Icons.alarm),
              //添加文本
              title: Text('Alarm'),
            ),
            ListTile(
              leading: Icon(Icons.phone),
              title: Text('Phone'),
            ),
            ListTile(
              leading: Icon(Icons.airplay),
              title: Text('Airplay'),
            ),
            ListTile(
              leading: Icon(Icons.airplay),
              title: Text('Airplay'),
            ),
            ListTile(
              leading: Icon(Icons.airplay),
              title: Text('Airplay'),
            ),
            ListTile(
              leading: Icon(Icons.airplay),
              title: Text('Airplay'),
            ),
            ListTile(
              leading: Icon(Icons.alarm),
              title: Text('Alarm'),
            ),
            ListTile(
              leading: Icon(Icons.alarm),
              title: Text('Alarm'),
            ),
```

```
          ListTile(
            leading: Icon(Icons.alarm),
            title: Text('Alarm'),
          ),
          ListTile(
            leading: Icon(Icons.alarm),
            title: Text('Alarm'),
          ),
          ListTile(
            leading: Icon(Icons.alarm),
            title: Text('Alarm'),
          ),
        ],
      ),
    );
  }
}
```

上述示例代码视图展现大致如图 4-8 所示。

4.5.2 水平列表组件

水平列表组件即为水平方向排列的组件,列表内部元素以水平方向排。把 ListView 组件的 scrollDirection 属性设置为 Axis.horizontal 即可。

示例代码如下:

图 4-8 ListView 组件应用示例

```
import 'package:flutter/material.dart';

void main() => runApp(new MyApp());

class MyApp extends StatelessWidget {
  @override
  Widget build(BuildContext context) {
    final title = "水平列表示例";

    return MaterialApp(
      title: title,
      home: Scaffold(
        appBar: AppBar(
          title: Text(title),
        ),
        body: Container(
          margin: EdgeInsets.symmetric(vertical: 20.0),
          height: 200.0,
          child: ListView(
```

```
            // 设置水平方向排列
            scrollDirection: Axis.horizontal,
            // 添加子元素
            children: <Widget>[
              Container(
                width: 160.0,
                color: Colors.lightBlue,
              ),
              Container(
                width: 160.0,
                color: Colors.amber,
              ),
              Container(
                width: 160.0,
                color: Colors.green,
                child: Column(
                  children: <Widget>[
                    Text(
                      '水平',
                      style: TextStyle(
                        fontWeight: FontWeight.bold,
                        fontSize: 36.0,
                      ),
                    ),
                    Text(
                      '列表',
                      style: TextStyle(
                        fontWeight: FontWeight.bold,
                        fontSize: 36.0,
                      ),
                    ),
                    Icon(Icons.list),
                  ],
                ),
              ),
              Container(
                width: 160.0,
                color: Colors.deepPurpleAccent,
              ),
              Container(
                width: 160.0,
                color: Colors.black,
              ),
              Container(
                width: 160.0,
                color: Colors.pinkAccent,
              ),
            ],
          ),
        ),
```

```
        ),
    );
  }
}
```

上述示例代码视图展现大致如图4-9所示。

4.5.3 长列表组件

当列表的数据项非常多时，需要使用长列表，比如淘宝后台订单列表、手机通讯录等，这些列表项数据很多。长列表也是使用ListView作为基础组件，只不过需要添加一个列表项构造器itemBuilder。

接下来通过一个示例来说明itemBuilder的用法，代码如下：

图4-9 水平列表组件应用示例

```
import 'package:flutter/material.dart';

void main() => runApp(new MyApp(
  // 使用generate方法产生500条数据
  items: new List<String>.generate(500, (i) =>
    "Item $i"),
));

class MyApp extends StatelessWidget {

  // 列表数据集
  final List<String> items;

  MyApp({Key key, @required this.items}) : super(key: key);

  @override
  Widget build(BuildContext context) {
    final title = "长列表示例";

    return MaterialApp(
      title: title,
      home: new Scaffold(
        appBar: new AppBar(
          title: new Text(title),
        ),
        // 使用ListView.builder来构造列表项
        body: new ListView.builder(
          // 列表长度
          itemCount: items.length,
          // 列表项构造器
          itemBuilder: (context,index) {
```

```
      return new ListTile(
        leading: new Icon(Icons.phone),
        title: new Text('${items[index]}'),
      );
    },
  ),
 ),
);
}
}
```

上述示例代码视图展现大致如图 4-10 所示。

4.5.4 网格列表组件

数据量很大时用矩阵方式排列比较清晰,此时用网格列表组件,即为 GridView 组件,可以实现多行多列的应用场景。使用 GridView 创建网格列表有多种方式:

❑ GridView.count 通过单行展示个数创建 GridView。
❑ GridView.extent 通过最大宽度创建 GridView。

网格列表组件的属性参见表 4-9。

接下来,我们选取 GridView.count 来创建风格列表,示例代码如下:

图 4-10 长列表组件应用示例

表 4-9 GridView 组件属性及描述

属性名	类型	默认值	说明
scrollDirection	Axis	Axis.vertical	滚动的方向,有垂直和水平两种,默认为垂直方向
reverse	bool	false	默认是从上或者左向下或者右滚动的,这个属性控制是否反向,默认值为 false,即不反向滚动
controller	ScrollController		控制 child 滚动时候的位置
primary	bool		是否是与父节点的 PrimaryScrollController 所关联的主滚动视图
physics	ScrollPhysics		滚动的视图如何响应用户的输入
shrinkWrap	bool	false	滚动方向的滚动视图内容是否应该由正在查看的内容所决定
padding	EdgeInsetsGeometry		四周的空白区域
gridDelegate	SliverGridDelegate		控制 GridView 中子节点布局的 delegate
cacheExtent	double		缓存区域

```
import 'package:flutter/material.dart';

void main() => runApp(new MyApp());

class MyApp extends StatelessWidget {
  @override
```

```
Widget build(BuildContext context) {
  final title = "网格列表示例";

  return new MaterialApp(
    title: title,
    home: new Scaffold(
      appBar: new AppBar(
        title: new Text(title),
      ),
      // 使用 GridView.count 构建网格
      body: new GridView.count(
        primary: false,
        // 四周增加一定的空隙
        padding: const EdgeInsets.all(20.0),
        crossAxisSpacing: 30.0,
        // 一行上放三列数据
        crossAxisCount: 3,
        // 数据项 五行三列
        children: <Widget>[
          const Text('第一行第一列'),
          const Text('第一行第二列'),
          const Text('第一行第三列'),
          const Text('第二行第一列'),
          const Text('第二行第二列'),
          const Text('第二行第三列'),
          const Text('第三行第一列'),
          const Text('第三行第二列'),
          const Text('第三行第三列'),
          const Text('第二行第三列'),
          const Text('第三行第一列'),
          const Text('第三行第二列'),
          const Text('第三行第一列'),
          const Text('第三行第二列'),
          const Text('第三行第三列'),
        ],
      ),
    ),
  );
}
```

上述示例代码视图展现大致如图 4-11 所示。

图 4-11 GridView 列表组件应用示例

4.6 表单组件

表单是一个包含表单元素的区域。表单元素允许用户输入内容，比如：文本域、下拉列表、单选框、复选框等。常见的应用场景有：登录、注册、输入信息等。表单里有两个重要的组件，一个是 Form 组件用来做整个表单提交使用的，另一个是 TextFormField 组件用来做用户输入的。

先来看看 Form 组件的属性，如下所示：

属性名	类型	说明
key	Key	组件在整个 Widget 树中的 key 值。
autovalidate	bool	是否自动提交表单。
child	Widget	组件 child 只能有一个子组件。
onChanged	VoidCallback	当 FormFiled 值改变时的回调函数。

再来看看 TextFromField 组件的属性，如下所示：

属性名	类型	说明
autovalidate	bool	自动验证值。
initialValue	T	表单字段初始值，比如：输入收货地址时，默认回填本的地址信息。
onSaved	FormFieldSetter<T>	当 Form 表单调用保存方法 Save 时回调的函数。
validator	FormFieldValidator<T>	Form 表单验证器。

对于输入框我们最关心的是输入内容是否合法，比如邮箱地址是否正确，电话号码是否是数字，等等。等用户输入完成后，需要知道输入框输入的内容。

为了获取表单的实例，我们需要设置一个全局类型的 key，通过这个 key 的属性，来获取表单对象。需要使用 GlobalKey 来获取，代码如下所示：

```
GlobalKey<FormState> loginKey = new GlobalKey<FormState>();
```

接下来编写一个简单的登录界面，其中有用户名，密码输入框再加上一个登录按钮。当用户没有输入任何内容时，输入提示"请输入用户名 请输入密码"，如图 4-12 所示。当用户正常输入内容点击"登录"按钮，界面不发生任何变化，表示输入内容正常。这里可以尝试输入一个非法的数据，比如验证代码里要求密码必需为 6 位，当你输入 6 位以下时，再次点击"登录"，会报一个错误提示信息"密码长度不够 6 位"，如图 4-13 所示。

完整示例代码如下所示：

```
import 'package:flutter/material.dart';

void main() => runApp(new LoginPage());

class LoginPage extends StatefulWidget {
  @override
  _LoginPageState createState() => new _LoginPageState();
}

class _LoginPageState extends State<LoginPage> {
  // 全局 Key 用来获取 Form 表单组件
  GlobalKey<FormState> loginKey = new GlobalKey<FormState>();

  // 用户名
```

```dart
String userName;

// 密码
String password;

void login() {
  // 读取当前的 Form 状态
  var loginForm = loginKey.currentState;

  // 验证 Form 表单
  if (loginForm.validate()) {
    loginForm.save();
    print('userName:' + userName + ' password:' + password);
  }
}

@override
Widget build(BuildContext context) {
  return new MaterialApp(
    title: 'Form 表单示例',
    home: new Scaffold(
      appBar: new AppBar(
        title: new Text('Form 表单示例'),
      ),
      body: new Column(
        children: <Widget>[
          new Container(
            padding: const EdgeInsets.all(16.0),
            child: new Form(
              key: loginKey,
              child: new Column(
                children: <Widget>[
                  new TextFormField(

                    decoration: new InputDecoration(
                      labelText: '请输入用户名',
                    ),
                    onSaved: (value) {
                      userName = value;
                    },
                    onFieldSubmitted: (value){

                    },
                  ),
                  new TextFormField(
                    decoration: new InputDecoration(
                      labelText: '请输入密码',
                    ),
                    obscureText: true,
                    // 验证表单方法
                    validator: (value) {
                      return value.length < 6 ? "密码长度不够6位" : null;
                    },
```

```
                onSaved: (value) {
                  password = value;
                },
              ),
            ],
          ),
        ),
      ),
      new SizedBox(
        width: 340.0,
        height: 42.0,
        child: new RaisedButton(
          onPressed: login,
          child: new Text(
            '登录',
            style: TextStyle(
              fontSize: 18.0,
            ),
          ),
        ),
      ),
    ],
   ),
  ),
 );
}
}
```

图 4-12　表单示例初始状态

图 4-13　表单示例错误状态

Chapter 5 第 5 章

Material Design 风格组件

Material Design 是由 Google 推出的全新设计语言,这种设计语言旨在为手机、平板电脑、台式机和其他平台提供更一致、更广泛的外观和感觉。在本书里我们有时把 Material Design 也称为纸墨设计,Material Design 风格是一种非常有质感的设计风格,并会提供一些默认的交互动画。

主要的 Material Design 风格组件参见表 5-1,本章将按照以下分类介绍这些组件:

❏ App 结构和导航组件
❏ 按钮和提示组件
❏ 其他组件

表 5-1 Material Design 风格组件的说明

组件名称	中文名称	简单说明
AppBar	应用按钮组件	应用的工具按钮
AlertDialog	对话框组件	有操作按钮的对话框
BottomNavigationBar	底部导航条组件	底部导航条,可以很容易地在 tap 之间切换和浏览顶级视图
Card	卡片组件	带有边框阴影的卡片组件
Drawer	抽屉组件	Drawer 抽屉组件可以实现类似抽屉拉开关闭的效果
FloatingActionButton	浮动按钮组件	应用的主要功能操作按钮
FlatButton	扁平按钮组件	扁平化风格的按钮
MaterialApp	Material 应用组件	MaterialApp 代表使用纸墨设计风格的应用
PopupMenuButton	弹出菜单组件	弹出菜单按钮
Scaffold	脚手架组件	实现了基本的 Material Design 布局
SnackBar	轻量提示组件	一个轻量级消息提示组件,在屏幕的底部显示
SimpleDialog	简单对话框组件	简单对话框组件,只起提示作用,没有交互

(续)

组件名称	中文名称	简单说明
TabBar	水平选项卡及视图组件	一个显示水平选项卡的 Material Design 组件
TextField	文本框组件	可接受应用输入文本的组件

5.1 App 结构和导航组件

本节介绍的这类组件对 App 的结构和导航却有帮助,如 MaterialApp、Scaffold、AppBar、BattomNavigationBar、TabBar、Drawer 等。

5.1.1 MaterialApp(应用组件)

MaterialApp 代表使用纸墨设计风格的应用,里面包含了其所需要的基本控件。一个完整的 Flutter 项目就是从 MaterialApp 这个主组件开始的。

MaterialApp 组件常见属性见表 5-2。

表 5-2 MaterialApp 组件属性及描述

属性名	类型	说明
title	String	应用程序的标题。该标题出现在如下位置: ● Android:任务管理器的程序快照之上 ● IOS:程序切换管理器中
theme	ThemeData	定义应用所使用的主题颜色,可以指定一个主题中每个控件的颜色
color	Color	应用的主要颜色值,即 primary color
home	Widget	这个是一个 Widget 对象,用来定义当前应用打开时,所显示的界面
routes	Map<String, WidgetBuilder>	定义应用中页面跳转规则
initialRoute	String	初始化路由
onGenerateRoute	RouteFactory	路由回调函数。当通过 Navigator.of(context).pushNamed 跳转路由时,在 routes 查找不到时,会调用该方法
onLocaleChanged		当系统修改语言的时候,会触发这个回调
navigatorObservers	List<NavigatorObserver>	导航观察器
debugShowMaterialGrid	bool	是否显示纸墨设计基础布局网格,用来调试 UI 的工具
showPerformanceOverlay	bool	显示性能标签

1. 设置主页

使用 home 属性设置应用的主页,即整个应用的主组件。示例代码如下:

```
import 'package:flutter/material.dart';

void main() {
  runApp(new MyApp());
```

```
}

class MyApp extends StatelessWidget {
  // 这是整个应用的主组件
  @override
  Widget build(BuildContext context) {
    return new MaterialApp(
      home: new MyHomePage(),
      title: 'MaterialApp 示例',
    );
  }
}
// 这是一个可改变的 Widget
class MyHomePage extends StatefulWidget {
  @override
  _MyHomePageState createState() => new _MyHomePageState();
}
class _MyHomePageState extends State<MyHomePage> {
  @override
  Widget build(BuildContext context) {
    return new Scaffold(
      appBar: new AppBar(
        title: Text('MaterialApp 示例'),
      ),
      body: Center(
        child: Text('主页'),
      ),
    );
  }
}
```

上述示例代码的视图展现大致如图 5-1 所示。

2. 路由处理

routes 对象是一个 Map<String，WidgetBuilder>。当使用 Navigator.pushNamed 来路由的时候，会在 routes 查找路由名字，然后使用对应的 WidgetBuilder 来构造一个带有页面切换动画的 MaterialPageRoute。如果应用只有一个界面，则不用设置这个属性，使用 home 设置这个界面即可。

图 5-1　MaterialApp 设置主页示例

通过 routes 可以给 MaterialApp 组件初始化一个路由列表，跳转到指定页面，代码如下所示：

```
Navigator.pushNamed(context, '/somePage');
```

在 MaterialApp 组件使用 initialRoute 属性可以给应用添加一个初始化路由。这两个属

性的代码如下:

```
routes: {
'/first': (BuildContext context) => FirstPage(), // 添加路由
'/second': (BuildContext context) => SecondPage(),
},
initialRoute: '/first',// 初始路由页面为 first 页面
```

为了便于演示,这里我们需要添加两个页面,在第一个页面里添加一个 Button,当点击 Button 时路由到第二个页面。页面跳转代码如下:

```
Navigator.pushNamed(context, '/second');
```

在第二个页面里也添加一个 Button,当点击 Button 时路由到第一个页面。页面跳转代码如下:

```
Navigator.pushNamed(context, '/first');
```

注意　第一个页面和第二个页的路由标识是 /first 及 /second 而不是 first 及 second。

完整的路由示例代码如下:

```
import 'package:flutter/material.dart';

void main() {
  runApp(new MyApp());
}

class MyApp extends StatelessWidget {
  // 这是整个应用的主组件
  @override
  Widget build(BuildContext context) {
    return new MaterialApp(
      home: new MyHomePage(),
      title: 'MaterialApp 示例',
      routes: {
        '/first': (BuildContext context) => FirstPage(), // 添加路由
        '/second': (BuildContext context) => SecondPage(),
      },
      initialRoute: '/first',// 初始路由页面为 first 页面
    );
  }
}

// 这是一个可改变的 Widget
class MyHomePage extends StatefulWidget {
  @override
  _MyHomePageState createState() => new _MyHomePageState();
```

```dart
  }

class _MyHomePageState extends State<MyHomePage> {
  @override
  Widget build(BuildContext context) {
    return new Scaffold(
      appBar: new AppBar(
        title: Text('MaterialApp 示例'),
      ),
      body: Center(
        child: Text(
          '主页',
          style: TextStyle(fontSize: 28.0),
        ),
      ),
    );
  }
}

// 第一个路由页面
class FirstPage extends StatelessWidget {
  @override
  Widget build(BuildContext context) {
    return new Scaffold(
      appBar: new AppBar(
        title: Text('这是第一页'),
      ),
      body: Center(
        child: RaisedButton(
          onPressed: () {
            // 路由跳转到第二个页面
            Navigator.pushNamed(context, '/second');
          },
          child: Text(
            '这是第一页',
            style: TextStyle(fontSize: 28.0),
          ),
        ),
      ),
    );
  }
}

// 第二个路由页面
class SecondPage extends StatelessWidget {
  @override
  Widget build(BuildContext context) {
    return new Scaffold(
      appBar: new AppBar(
        title: Text('这是第二页'),
```

```
      ),
      body: Center(
        child: RaisedButton(
          onPressed: () {
            // 路由跳转到第一个页面
            Navigator.pushNamed(context, '/first');
          },
          child: Text(
            '这是第二页',
            style: TextStyle(fontSize: 28.0),
          ),
        ),
      ),
    );
  }
}
```

上述示例代码的视图展现大致如图 5-2 所示。

图 5-2　MaterialApp 路由处理示例

3. 自定义主题

应用程序的主题，各种定制的颜色都可以设置，用于程序主题切换。示例代码如下所示：

```
new MaterialApp(
    theme: new ThemeData(
        // 主题色
```

```
      primarySwatch: Colors.blue,
    ),
);
```

5.1.2 Scaffold（脚手架组件）

Scaffold 实现了基本的 Material Design 布局。只要是在 Material Design 中定义过的单个界面显示的布局组件元素，都可以使用 Scaffold 来绘制。

Scaffold 组件常见属性如表 5-3 所示。

表 5-3 Scaffold 组件属性及描述

属性名	类型	说明
appBar	AppBar	显示在界面顶部的一个 AppBar
body	Widget	当前界面所显示的主要内容
floatingActionButton	Widget	在 Material Design 中定义的一个功能按钮
persistentFooterButtons	List<Widget>	固定在下方显示的按钮
drawer	Widget	侧边栏组件
bottomNavigationBar	Widget	显示在底部的导航栏按钮栏
backgroundColor	Color	背景颜色
resizeToAvoidBottomPadding	bool	控制界面内容 body 是否重新布局来避免底部被覆盖，比如当键盘显示时，重新布局避免被键盘盖住内容。默认值为 true

示例代码如下：

```
import 'package:flutter/material.dart';

void main() => runApp(
  new MaterialApp(
    title: 'Scaffold 脚手架组件示例',
    home: new LayoutDemo(),
  ),
);

class LayoutDemo extends StatelessWidget {

  @override
  Widget build(BuildContext context) {

    return new Scaffold(
      // 头部元素 比如：左侧返回按钮 中间标题 右侧菜单
      appBar: AppBar(
        title: Text('Scaffold 脚手架组件示例'),
      ),
      // 视图内容部分
      body: Center(
        child: Text('Scaffold'),
```

```
    ),
    //底部导航栏
    bottomNavigationBar: BottomAppBar(
      child: Container(height: 50.0,),
    ),
    // 添加 FAB 按钮
    floatingActionButton: FloatingActionButton(
      onPressed: () {},
      tooltip: '增加',
      child: Icon(Icons.add),
    ),
    //FAB 按钮居中展示
    floatingActionButtonLocation: Floating
      ActionButtonLocation.centerDocked,
    );
  }
}
```

上述示例代码的视图展现大致如图 5-3 所示。

5.1.3 AppBar（应用按钮组件）

应用按钮组件有 AppBar 和 SliverAppBar。它们是纸墨设计中的 AppBar，也就是 Android 中的 Toolbar。

AppBar 和 SliverAppBar 都是继承自 StatefulWidget 类，都代表 Toolbar，两者的区别在于 AppBar 位置是固定在应用最上面的；而 SliverAppBar 是可以跟随内容滚动的。AppBar 及 SliverAppBar 组件常见属性见表 5-4。

图 5-3 Scaffold 抽屉组件示例

表 5-4 AppBar 及 SliverAppBar 组件属性及描述

属性名	类型	默认值	说明
leading	Widget	null	在标题前面显示的一个组件，在首页通常显示应用的 logo；在其他界面通常显示为返回按钮
title	Widget	null	Toolbar 中主要内容，通常显示为当前界面的标题文字
actions	List<Widget>	null	一个 Widget 列表，代表 Toolbar 中所显示的菜单，对于常用的菜单，通常使用 IconButton 来表示，对于不常用的菜单通常使用 PopupMenuButton 来显示为三个点，点击后弹出二级菜单
bottom	PreferredSizeWidget	null	通常是 TabBar。用来在 Toolbar 标题下面显示一个 Tab 导航栏
elevation	double	4	纸墨设计中组件的 z 坐标顺序，对于可滚动的 SliverAppBar，当 SliverAppBar 和内容同级的时候，该值为 0，当内容滚动 SliverAppBar 变为 Toolbar 的时候，修改 elevation 的值

（续）

属性名	类型	默认值	说明
flexibleSpace	Widget	null	一个显示在 AppBar 下方的组件，高度和 AppBar 高度一样，可以实现一些特殊的效果，该属性通常在 SliverAppBar 中使用
backgroundColor	Color	ThemeData.primaryColor	背景色
brightness	Brightness	ThemeData.primaryColorBrightness	AppBar 的亮度，有白色和黑色两种主题
iconTheme	IconThemeData	ThemeData.primaryIconTheme	AppBar 上图标的颜色、透明度和尺寸信息。默认值为 ThemeData.primaryIconTheme
textTheme	TextTheme	ThemeData.primaryTextTheme	AppBar 上的文字样式
centerTitle	bool	true	标题是否居中显示，默认值根据不同的操作系统，显示方式不一样

AppBar 可以显示顶部 leading、title 和 actions 等内容。底部通常为选项卡 TabBar。flexibleSpace 显示在 AppBar 的下方，高度和 AppBar 高度一样，可以实现一些特殊的效果，不过该属性通常在 SliverAppBar 中使用。具体布局如图 5-4 所示。

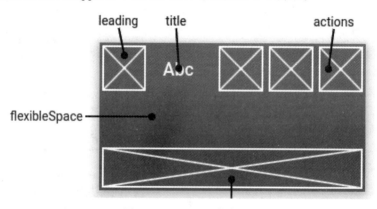

图 5-4　AppBar 组件布局图

接下来是一个示例代码，在上端左侧显示标题，右侧添加两个按钮，示例代码如下：

```
import 'package:flutter/material.dart';

void main() => runApp(
  new MaterialApp(
    title: 'AppBar 应用按钮示例',
    home: new LayoutDemo(),
  ),
);

class LayoutDemo extends StatelessWidget {
```

```
@override
Widget build(BuildContext context) {

  return new Scaffold(
    appBar: AppBar(
      title: Text('AppBar 应用按钮示例'),
      actions: <Widget>[
        IconButton(
          icon: Icon(Icons.search),
          tooltip: '搜索',
          onPressed: (){},
        ),
        IconButton(
          icon: Icon(Icons.add),
          tooltip: '添加',
          onPressed: (){},
        ),
      ],
    ),
  );
}
```

上述示例代码的视图展现大致如图 5-5 所示。

图 5-5　AppBar 应用按钮组件示例

5.1.4　BottomNavigationBar（底部导航条组件）

BottomNavigationBar 是底部导航条，可以很容易地在 tap 之间切换和浏览顶级视图。很多 App 主页底部都采用这种切换的方式。

BottomNavigationBar 组件常见属性见表 5-5。

表 5-5　BottomNavigationBar 组件属性及描述

属性名	类型	说明
currentIndex	int	当前索引，用来切换按钮控制
fixedColor	Color	选中按钮的颜色。如果没有指定值，则用系统主题色
iconSize	double	按钮图标大小
items	List<BottomNavigationBarItem>	底部导航条按钮集。每一项是一个 BottomNavigationBarItem，有 icon 图标及 title 文本属性
onTap	ValueChanged<int>	按下其中某一个按钮回调事件。需要根据返回的索引设置当前索引

BottomNavigationBar 通常显示在应用页面底部。由按钮加文字组成，这个可以根据实际应用场景任意组合。按下按钮切换不同的页面。所以它需要一个当前索引来控制当前具体切换的页面。布局示例如图 5-6 所示。

图 5-6　BottomNavigationBar 组件布局图

接下来是一个示例代码，仿一个聊天软件，底部显示"信息"、"通讯录"及"发现"按钮。这个界面有点像微信的排列方式。按钮是由图标加文本方式，切换不同的按钮，中间显示切换的内容。示例中用到了状态控制，用来控制当前切换按钮的索引值。

完整的示例代码如下：

```dart
import 'package:flutter/material.dart';

void main() => runApp(MyApp());

class MyApp extends StatelessWidget {
  @override
  Widget build(BuildContext context) {
    return MaterialApp(
      home: Scaffold(
        body: MyHomePage(),
      ),
    );
  }
}

class MyHomePage extends StatefulWidget {
  MyHomePage({Key key}) : super(key: key);

  @override
  _MyHomePageState createState() => _MyHomePageState();
}

class _MyHomePageState extends State<MyHomePage> {
  int _selectedIndex = 1;// 当前选中项的索引
  final _widgetOptions = [
    Text('Index 0: 信息'),
```

```
    Text('Index 1：通讯录'),
    Text('Index 2：发现'),
];

@override
Widget build(BuildContext context) {
    return Scaffold(
      appBar: AppBar(
        title: Text('BottomNavigationBar 示例'),
      ),
      body: Center(
        child: _widgetOptions.elementAt(_selectedIndex),// 居中显示某一个文本
      ),
      // 底部导航按钮 包含图标及文本
      bottomNavigationBar: BottomNavigationBar(
        items: <BottomNavigationBarItem>[
          BottomNavigationBarItem(icon: Icon(Icons.chat), title: Text('信息')),
          BottomNavigationBarItem(icon: Icon(Icons.contacts), title: Text('通讯录')),
          BottomNavigationBarItem(icon: Icon(Icons.account_circle), title: Text('发现')),
        ],
        currentIndex: _selectedIndex,// 当前选中项的索引
        fixedColor: Colors.deepPurple,// 选项中项的颜色
        onTap: _onItemTapped,// 选择按下处理
      ),
    );
}

// 选择按下处理 设置当前索引为 index 值
void _onItemTapped(int index) {
    setState(() {
      _selectedIndex = index;
    });
}
}
```

上述示例代码的视图展现大致如图 5-7 所示。

5.1.5 TabBar（水平选项卡及视图组件）

TabBar 是一个显示水平选项卡的 Material Design 组件，通常需要配套 Tab 选项组件及 TabBarView 页面视图组件一起使用。

图 5-7 BottomNavigationBar 组件示例

TabBar 组件常见属性如下所示：

属性名	类型	说明
isScrollable	bool	是否可以水平移动。
tabs	List<Widget>	Tab 选项列表，建议不要放太多项，否则用户操作起来不方便。

Tab 组件常见属性如下所示：

属性名	类型	说明
icon	Widget	Tab 图标。
text	String	Tab 文本。

TabBarView 组件常见属性如下所示：

属性名	类型	说明
controller	TabController	指定视图的控制器。
children	List<Widget>	视图组件的 child 为一个列表，一个选项卡对应一个视图。

TabBar 的选项卡按钮通常显示在应用页面上部，TabBarView 视图布局示例参考图 5-8。

相信大家都用过地图类软件。当你选择好目的地以后，它会给你列出几个选项，比如步行、自驾、公交车等，并给出需要多长时间以及对应路线之类的信息。接下来我们就用 TabBar 来实现一个类似的功能页面。

实现这个示例需要以下几个组件：

❑ DefaultTabController

❑ TabBar

❑ Tab

❑ TabBarView

图 5-8　TabBar 组件布局图

在编写示例之前，我们先了解一下 DefaultTabController 这个组件，它是 TabBar 和 TabBarView 的控制器，是关联这两组件的桥梁。

DefaultTabController 的示例代码如下：

```
import 'package:flutter/material.dart';


void main() {
  runApp(new DefaultTabControllerSample());
}

class DefaultTabControllerSample extends StatelessWidget {

  // 选项卡数据
  final List<Tab> myTabs = <Tab>[
    Tab(text: '选项卡一'),
    Tab(text: '选项卡二'),
  ];

  @override
```

```
Widget build(BuildContext context) {
  return new MaterialApp(
    // 用来组装 TabBar 及 TabBarView
    home: DefaultTabController(
      length: myTabs.length,
      child: Scaffold(
        appBar: AppBar(
          // 添加导航栏
          bottom: TabBar(
            tabs: myTabs,
          ),
        ),
        // 添加导航视图
        body: TabBarView(
          children: myTabs.map((Tab tab) {
            return Center(child: Text(tab.text));
          }).toList(),
        ),
      ),
    ),
  );
}
```

上述示例代码的视图展现大致如图 5-9 所示。

接下来我们编写 TabBar 水平选项卡的整个示例。步骤如下：

步骤 1：首先添加视图项数据 ItemView，增加 title 及 icon 两个数据项。结构如下所示：

```
class ItemView {
  title;// 标题
  icon;// 图标
}
```

步骤 2：组装数据，添加所有选项类目。以自驾为例代码如下所示。添加自驾、自行车、轮船、公交车、火车及步行数据：

图 5-9　DefaultTabController 组件示例

```
ItemView(title: '自驾', icon: Icons.directions_car),
```

最后用 List<ItemView> 组装起来：

```
const List<ItemView> items = const <ItemView>[];
```

步骤 3：添加被选中的视图命名为 SelectedView，此组件需要传入视图项数据，包含图标及标题，然后通过 Column 组件居中显示在界面里。代码结构如下所示：

```
// 被选中的视图
class SelectedView extends StatelessWidget {
  //TODO 构造方法

  // 视图数据
  final ItemView item;

  @override
  Widget build(BuildContext context) {
    // 渲染页面 添加图标和文本
  }
}
```

步骤 4：接着添加示例的主体命名为 TabBarSample。主要用 DefaultTabController 来包裹并指定选项卡的数量。接着在 AppBar 里添加 TabBar 组件。注意这里的 bottom 不要理解为底部，应理解为 AppBar 的按钮部分。TabBar 需要设置为可滚动的，因为选项卡子项较多，界面需要滑动才可以显示全。TabBar 的子项 tabs 需要使用迭代器来展示 Tab 内容。添加好 TabBar 后再添加视图 TabBarView，这里同样也需要使用迭代器来展示选项卡视图。迭代器代码如下所示：

```
tabs: items.map((ItemView item) {
    return new Tab(
          text: item.title,
            icon: new Icon(item.icon),
            );
}).toList(),
```

完整的示例代码如下所示：

```
import 'package:flutter/material.dart';

class TabBarSample extends StatelessWidget {
  @override
  Widget build(BuildContext context) {
    return new MaterialApp(
       // 添加 DefaultTabController 关联 TabBar 及 TabBarView
      home: new DefaultTabController(
        length: items.length,// 传入选项卡数量
        child: new Scaffold(
          appBar: new AppBar(
            title: const Text('TabBar 选项卡示例 '),
            bottom: new TabBar(
              isScrollable: true,// 设置为可以滚动
              tabs: items.map((ItemView item) {// 迭代添加选项卡子项
                return new Tab(
                  text: item.title,
                  icon: new Icon(item.icon),
                );
```

```
          }).toList(),
        ),
      ),
      // 添加选项卡视图
      body: new TabBarView(
        children: items.map((ItemView item) {// 迭代显示选项卡视图
          return new Padding(
            padding: const EdgeInsets.all(16.0),
            child: new SelectedView(item: item),
          );
        }).toList(),
      ),
    ),
  );
}

// 视图项数据
class ItemView {
  const ItemView({ this.title, this.icon });// 构造方法
  final String title;// 标题
  final IconData icon;// 图标
}

// 选项卡的类目
const List<ItemView> items = const <ItemView>[
  const ItemView(title: '自驾', icon: Icons.directions_car),
  const ItemView(title: '自行车', icon: Icons.directions_bike),
  const ItemView(title: '轮船', icon: Icons.directions_boat),
  const ItemView(title: '公交车', icon: Icons.directions_bus),
  const ItemView(title: '火车', icon: Icons.directions_railway),
  const ItemView(title: '步行', icon: Icons.directions_walk),
];

// 被选中的视图
class SelectedView extends StatelessWidget {
  const SelectedView({ Key key, this.item }) : super(key: key);

  // 视图数据
  final ItemView item;

  @override
  Widget build(BuildContext context) {
    final TextStyle textStyle = Theme.of(context).textTheme.display1;
    return new Card(
      color: Colors.white,
      child: new Center(
        child: new Column(
          mainAxisSize: MainAxisSize.min,// 垂直方向最小化处理
```

```
        crossAxisAlignment: CrossAxisAlignment.center,// 水平方向居中对齐
        children: <Widget>[
          new Icon(item.icon, size: 128.0, color: textStyle.color),
          new Text(item.title, style: textStyle),
        ],
      ),
    ),
  );
  }
}

void main() {
  runApp(new TabBarSample());
}
```

上述示例代码的视图展现大致如图 5-10 所示。

图 5-10　TabBar 选项卡示例

> 提示　选项卡项较多时，需要左右滑动才可以展示全。建议在做实际项目时，可以减少一些选项，尽量把更多的内容归在一起。这样可以提高用户体验效果。

5.1.6　Drawer（抽屉组件）

Drawer 可以实现类似抽屉拉出推入的效果。可以从侧边栏拉出导航面板。这样做的好

处是，可以把一些功能菜单折叠起来。通常 Drawer 是和 ListView 组件组合使用的。Drawer 组件常见属性如下所示：

属性名	类型	默认值	说明
child	Widget		Drawer 的 child 可以放置任意可显示对象。
elevation	double	16	纸墨设计中组件的 z 坐标顺序。

Drawer 组件可以添加头部效果，用以下两个组件可以实现：
❑ DrawerHeader：展示基本信息。
❑ UserAccountsDrawerHeader：展示用户头像、用户名、Email 等信息。
DrawerHeader 通常用于抽屉中在顶部展示一些基本信息，属性及描述如表 5-6 所示。

表 5-6　DrawerHeader 组件属性及描述

属性名	类型	说明
decoration	Decoration	header 区域的 decoration，通常用来设置背景颜色或者背景图片
curve	Curve	如果 decoration 发生了变化，则会使用 curve 设置的变化曲线和 duration 设置的动画时间来做一个切换动画
child	Widget	Header 里面所显示的内容控件
padding	EdgeInsetsGeometry	Header 里面内容控件的 padding 值，如果 child 为 null，则这个值无效
margin	EdgeInsetsGeometry	Header 四周的间隙

UserAccountsDrawerHeader 可以设置用户头像、用户名和 Email 等信息，显示一个符合 Material Design 规范的 drawer header。其常用属性如表 5-7 所示。

表 5-7　UserAccountsDrawerHeader 组件属性及描述

属性名	类型	说明
margin	EdgeInsetsGeometry	Header 四周的间隙
decoration	Decoration	header 区域的 decoration，通常用来设置背景颜色或者背景图片
currentAccountPicture	Widget	用来设置当前用户的头像
otherAccountsPictures	List<Widget>	用来设置当前用户其他账号的头像
accountName	Widget	当前用户的名字
accountEmail	Widget	当前用户的 Email
onDetailsPressed	VoidCallback	当 accountName 或者 accountEmail 被点击的时候所触发的回调函数，可以用来显示其他额外的信息

从 UserAccountsDrawerHeader 的属性及描述来看，我们可以写一个示例来模仿 QQ 侧边导航栏的效果。示例中用到的图片，请参考第 4 章中"图片组件"一节即可。

示例代码如下：

```
import 'package:flutter/material.dart';

void main() => runApp(
```

```
    new MaterialApp(
      title: 'Drawer 抽屉组件示例 ',
      home: new LayoutDemo(),
    ),
);

class LayoutDemo extends StatelessWidget {

  @override
  Widget build(BuildContext context) {

    return new Scaffold(
      appBar: AppBar(
        title: Text('Drawer 抽屉组件示例 '),
      ),
      drawer: Drawer(
        child: ListView(
          children: <Widget>[
            // 设置用户信息 头像、用户名、Email 等
            UserAccountsDrawerHeader(
              accountName: new Text(
                " 玄微子 ",
              ),
              accountEmail: new Text(
                "xuanweizi@163.com",
              ),
              // 设置当前用户头像
              currentAccountPicture: new CircleAvatar(
                backgroundImage: new AssetImage("images/1.jpeg"),
              ),
              onDetailsPressed: () {},
              // 属性本来是用来设置当前用户的其他账号的头像 这里用来当 QQ 二维码图片展示
              otherAccountsPictures: <Widget>[
                new Container(
                  child: Image.asset('images/code.jpeg'),
                ),
              ],
            ),
            ListTile(
              leading: new CircleAvatar(child: Icon(Icons.color_lens)),// 导航栏菜单
              title: Text(' 个性装扮 '),
            ),
            ListTile(
              leading: new CircleAvatar(child: Icon(Icons.photo)),
              title: Text(' 我的相册 '),
            ),
            ListTile(
              leading: new CircleAvatar(child: Icon(Icons.wifi)),
              title: Text(' 免流量特权 '),
            ),
```

```
            ],
          ),
        ),
      );
  }
}
```

上述示例代码的视图展现大致如图 5-11 所示。

5.2 按钮和提示组件

本节介绍的组件可帮助设计按钮、菜单、对话框等，包括：FloatingActionButton、FlatButton、PopupMenuButton、SimpleDialog、AlertDialog、SnackBar。

5.2.1 FloatingActionButton（悬停按钮组件）

FloatingActionButton 对应一个圆形图标按钮，悬停在内容之上，以展示应用程序中的主要动作，所以非常醒目，类似于 iOS 系统里的小白点按钮。FloatingActionButton 通常用于 Scaffold.floatingActionButton 字段，常用属性见表 5-8。

图 5-11　Drawer 抽屉组件示例

表 5-8　FloatingActionButton 组件属性及描述

属性名	类型	默认值	说明
child	Widget		child 一般为 Icon，不推荐使用文字
tooltip	String		按钮提示文本
foregroundColor	Color		前景色
backgroundColor	Color		背景色
elevation	double	6.0	未点击时阴影值，默认 6.0
hignlightElevation	double	12.0	点击时阴影值
onPressed	VoidCallback		点击事件回调
shape	ShapeBorder	CircleBorder	定义按钮的 shape，设置 shape 时，默认的 elevation 将会失效，默认为 CircleBorder

示例代码如下：

```
import 'package:flutter/material.dart';

void main() => runApp(MyApp());

class MyApp extends StatelessWidget {
```

```
      @override
      Widget build(BuildContext context) {
        return MaterialApp(
          title: 'FloatingActionButton示例',
          home: Scaffold(
            appBar: AppBar(
              title: Text('FloatingActionButton示例'),
            ),
            body: Center(
              child: Text(
                'FloatingActionButton示例',
                style: TextStyle(fontSize: 28.0),
              ),
            ),
            floatingActionButton: new Builder(builder: (BuildContext context) {
              return new FloatingActionButton(
                child: const Icon(Icons.add),
                //提示信息
                tooltip: "请点击FloatingActionButton",
                //前景色为白色
                foregroundColor: Colors.white,
                //背景色为蓝色
                backgroundColor: Colors.blue,
                //未点击阴影值
                elevation: 7.0,
                //点击阴影值
                highlightElevation: 14.0,
                onPressed: () {
                  //点击回调事件 弹出一句提示语句
                  Scaffold.of(context).showSnackBar(new SnackBar(
                    content: new Text("你点击了FloatingActionButton"),
                  ));
                },
                mini: false,
                //圆形边
                shape: new CircleBorder(),
                isExtended: false,
              );
            }),
            floatingActionButtonLocation:
            FloatingActionButtonLocation.centerFloat, //居中放置 位置可以设置成左中右
          ),
        );
      }
    }
```

上述示例代码的视图展现大致如图5-12所示，当按下图中按钮后，会触发onPressed事件，调用SnackBar显示一句提示的话。

图 5-12　FloatingActionButton 组件应用示例

5.2.2　FlatButton（扁平按钮组件）

FlatButton 组件是一个扁平的 Material Design 风格按钮，点击时会有一个阴影效果。示例代码如下：

```
import 'package:flutter/material.dart';

void main() => runApp(MyApp());

class MyApp extends StatelessWidget {
  @override
  Widget build(BuildContext context) {
    return MaterialApp(
      title: 'Welcome to Flutter',
      home: Scaffold(
        appBar: AppBar(
          title: Text('FlatButton 扁平按钮组件示例 '),
        ),
        body: Center(
          child: FlatButton(
            onPressed: () {},
            child: Text(
              'FlatButton',
```

```
              style: TextStyle(fontSize: 24.0),
            ),
          ),
        ),
      );
    }
  }
```

上述示例代码的视图展现大致如图 5-13 所示。

图 5-13　FlatButton 组件应用示例

5.2.3　PopupMenuButton（弹出菜单组件）

PopupMenuButton 为弹出菜单按钮，一般放在应用页面的右上角，表示有更多的操作。菜单项使用 PopupMenuItem 组件。常用属性见表 5-9。

表 5-9　PopupMenuButton 组件属性及描述

属性名	类型	说明
child	Widget	child 如果提供则弹出菜单组件将使用此组件
icon	Icon	如果提供，则弹出菜单使用此图标
itemBuilder	PopupMenuItemBuilder<T>	菜单项构造器，菜单项为任意类型，文本，图标都可以
onSelected	PopupMenuItemSelected<T>	当某项菜单被选中时回调的方法

接下来我们编写一个视频会议产品中关于会控的菜单。视频会议的会控菜单有如下内容：
- 添加成员
- 锁定会议
- 修改布局
- 挂断所有

完整的示例代码如下：

```dart
import 'package:flutter/material.dart';

void main() => runApp(MyApp());

// 会控菜单项
enum ConferenceItem { AddMember, LockConference, ModifyLayout, TurnoffAll }

class MyApp extends StatelessWidget {
  @override
  Widget build(BuildContext context) {
    return MaterialApp(
      title: 'PopupMenuButton 组件示例 ',
      home: Scaffold(
        appBar: AppBar(
          title: Text('PopupMenuButton 组件示例 '),
        ),
        body: Center(
          child: FlatButton(
            onPressed: () {},
            child: PopupMenuButton<ConferenceItem>(
              onSelected: (ConferenceItem result) {},
              itemBuilder: (BuildContext context) =>// 菜单项构造器
                  <PopupMenuEntry<ConferenceItem>>[
                const PopupMenuItem<ConferenceItem>(// 菜单项
                  value: ConferenceItem.AddMember,
                  child: Text(' 添加成员 '),
                ),
                const PopupMenuItem<ConferenceItem>(
                  value: ConferenceItem.LockConference,
                  child: Text(' 锁定会议 '),
                ),
                const PopupMenuItem<ConferenceItem>(
                  value: ConferenceItem.ModifyLayout,
                  child: Text(' 修改布局 '),
                ),
                const PopupMenuItem<ConferenceItem>(
                  value: ConferenceItem.TurnoffAll,
                  child: Text(' 挂断所有 '),
                ),
              ],
```

```
            ),
          ),
        ),
      ),
    );
  }
}
```

上述示例代码的视图展现大致如图 5-14 所示。

图 5-14　PopupMenuButton 组件应用示例

5.2.4　SimpleDialog（简单对话框组件）

SimpleDialog 组件用于设计简单对话框，可以显示附加的提示或操作，组件的属性及描述如表 5-10 所示。

表 5-10　SimpleDialog 组件属性及描述

属性名	类型	说明
children	List<Widget>	对话框子项，典型的应用是一个列表
title	Widget	通常是一个文本组件
contentPadding	EdgeInsetsGeometry	内容部分间距大小
titlePadding	EdgeInsetsGeometry	标题部分间距大小

简单对话框通常需要配合 SimpleDialogOption 组件一起使用，接下来通过一个示例来展示如何使用，示例代码如下：

```
import 'package:flutter/material.dart';

void main() => runApp(MyApp());

class MyApp extends StatelessWidget {
  @override
  Widget build(BuildContext context) {

    return MaterialApp(
      title: 'SimpleDialog 组件示例',
      home: Scaffold(
        appBar: AppBar(
          title: Text('SimpleDialog 组件示例'),
        ),
        body: Center(
          child: SimpleDialog(
            title: const Text('对话框标题'),
            children: <Widget>[
              SimpleDialogOption(
                onPressed: () { },
                child: const Text('第一行信息'),
              ),
              SimpleDialogOption(
                onPressed: () { },
                child: const Text('第二行信息'),
              ),
            ],
          ),
        ),
      ),
    );
  }
}
```

上述示例代码的视图展现大致如图 5-15 所示。

图 5-15　SimpleDialog 组件应用示例

> **注意**　通常对话框都是某个动作来触发渲染的。比如点击按钮，点击菜单等。所以对话框一般要封装在一个方法里实现。另外这个过程是异步的需要加入 async/await 处理。

5.2.5　AlertDialog（提示对话框组件）

AlertDialog 组件比 SimpleDialog 对话框又复杂一些。不仅仅有提示内容，还有一些操作按钮，如确定和取消等，内容部分可以用 SingleChildScrollView 进行包裹。组件的属性及描述如表 5-11 所示。

表 5-11　AlertDialog 组件属性及描述

属性名	类型	说明
actions	List<Widget>	对话框底部操作按钮，例如确定、取消等
title	Widget	通常是一个文本组件
contentPadding	EdgeInsetsGeometry	内容部分间距大小
content	Widget	内容部分，通常为对话框的提示内容
titlePadding	EdgeInsetsGeometry	标题部分间距大小

编写一个删除确认的示例，完整的示例代码如下：

```dart
import 'package:flutter/material.dart';

void main() => runApp(MyApp());

class MyApp extends StatelessWidget {
  @override
  Widget build(BuildContext context) {
    return MaterialApp(
      title: 'AlertDialog 组件示例 ',
      home: Scaffold(
        appBar: AppBar(
          title: Text('AlertDialog 组件示例 '),
        ),
        body: Center(
          child: AlertDialog(
            title: Text(' 提示 '), // 对话框标题
            content: SingleChildScrollView(
              // 对话框内容部分
              child: ListBody(
                children: <Widget>[
                  Text(' 是否要删除 ?'),
                  Text(' 一旦删除数据不可恢复！ '),
                ],
              ),
            ),
            actions: <Widget>[
              FlatButton(
                child: Text(' 确定 '),
                onPressed: () {},
              ),
              FlatButton(
                child: Text(' 取消 '),
                onPressed: () {},
              ),
            ],
          ),
        ),
```

```
            ),
          );
    }
}
```

上述示例代码的视图展现大致如图 5-16 所示。

图 5-16　AlertDialog 组件应用示例

5.2.6　SnackBar（轻量提示组件）

SnackBar 是一个轻量级消息提示组件，在屏幕的底部显示，组件的属性及描述如表 5-12 所示。

表 5-12　SnackBar 组件属性及描述

属性名	类型	默认值	说明
action	SnackBarAction		提示消息里执行的动作，比如用户想撤销时可以点击操作
animation	Animation<double>		给组件添加动画效果
content	Widget		提示消息内容，通常为文本组件
duration	Duration	4.0 秒	动画执行的时长
backgroundColor	Color		消息面板的背景色

弹出消息提示调用方法如下所示，过几秒钟会自动提示消息：

```
Scaffold.of(context).showSnackBar();
```

完整的示例代码如下：

```dart
import 'package:flutter/material.dart';

void main() => runApp(MyApp());

class MyApp extends StatelessWidget {
  @override
  Widget build(BuildContext context) {
    return MaterialApp(
      title: 'Welcome to Flutter',
      home: Scaffold(
        appBar: AppBar(
          title: Text('SnackBar 示例'),
        ),
        body: Center(
          child: Text(
            'SnackBar 示例',
            style: TextStyle(fontSize: 28.0),
          ),
        ),
        floatingActionButton: new Builder(builder: (BuildContext context) {
          return new FloatingActionButton(
            child: const Icon(Icons.add),
            onPressed: () {
              // 点击回调事件 弹出一句提示语句
              Scaffold.of(context).showSnackBar
                (new SnackBar(
                  content: new Text("显示SnackBar"),
                ));
            },
            shape: new CircleBorder(),
          );
        }),
        floatingActionButtonLocation:
        FloatingActionButtonLocation.endFloat,
        // 居中放置 位置可以设置成左中右
      ),
    );
  }
}
```

上述示例代码的视图展现大致如图 5-17 所示。

5.3 其他组件

本节介绍的组件包括文本框、卡片等内容的设计，如 TextField、Card 等。

图 5-17　SnackBar 组件应用示例

5.3.1　TextField（文本框组件）

只要是应用程序就少不了交互，文本输入是最常见的一种交互方式。TextField 组件就是用来做文本输入的组件。注意这个要和 Text 组件区分开来，Text 组件主要用于显示文本，并不接受输入文本。TextField 常用属性见表 5-13。

表 5-13　TextField 组件属性及描述

属性名	类型	说明
maxLength	int	最大长度
maxLines	int	最大行数
autocorrect	bool	是否自动更正
autofocus	bool	是否自动对焦
obscureText	bool	是否是密码
textAlign	TextAlign	文本对齐方式
style	TextStyle	输入文本的样式
inputFormatters	List<TextInputFormatter>	允许的输入格式
onChanged	ValueChanged<String>	内容改变的回调
onSubmitted	ValueChanged<String>	内容提交时回调
enabled	bool	是否禁用

TextField 最基本的用法就是什么都不处理，默认 Material Design 的效果就非常漂亮。代码如下所示：

```
TextField()
```

假设还想获取文本内容，仅有输入框还不行，还需要传递 controller 给 TextField。用来监听文本内容的变化，这是一种绑定的机制。初如化监听器代码如下所示：

```
final TextEditingController controller = TextEditingController();
controller.addListener(() {
    //TODO
});
```

绑定监听器代码如下所示：

```
child: TextField(
   controller: controller,
),
```

当你在文本框里输入"hello world"，你会发现在控制台里文字是一个一个打印出来的。原因是监听器只要发现文本内容有发生变化，就会触发回调函数输出内容。输出结果如下：

```
flutter: 你输入的内容为：
flutter: 你输入的内容为：h
flutter: 你输入的内容为：he
```

```
flutter: 你输入的内容为：hel
flutter: 你输入的内容为：hell
flutter: 你输入的内容为：hello
flutter: 你输入的内容为：hello
flutter: 你输入的内容为：hello w
flutter: 你输入的内容为：hello wo
flutter: 你输入的内容为：hello wor
flutter: 你输入的内容为：hello worl
flutter: 你输入的内容为：hello world
```

接下来可以再给 TextField 添加一些属性用来做更多的控制，比如最大长度，最大行数，是否自动对焦，内容提交回调等等。添加常规属性后效果如图 5-18 所示。

最后再给文本框加一些装饰效果，比如填充色，图标等内容。组件的装饰和生活中的装修一样任由个人发挥，这里建议简单大方即可。需要用到 InputDecoration 组件。代码如下所示：

```
decoration: InputDecoration
```

完整的示例代码如下：

```
import 'package:flutter/material.dart';

void main() => runApp(MyApp());

class MyApp extends StatelessWidget {
  @override
  Widget build(BuildContext context) {
    // 添加文本编辑控制器 监听文本输入内容变化
    final TextEditingController controller = TextEditingController();
    controller.addListener(() {
      print('你输入的内容为：${controller.text}');
    });

    return MaterialApp(
      title: 'TextField组件示例',
      home: Scaffold(
        appBar: AppBar(
          title: Text('TextField组件示例'),
        ),
        body: Center(
          child: Padding(
            padding: const EdgeInsets.all(20.0),
            child: TextField(
              // 绑定controller
              controller: controller,
```

图 5-18 TextField 添加常规属性示例

```
            // 最大长度，设置此项会让 TextField 右下角有一个输入数量的统计字符串
            maxLength: 30,
            // 最大行数
            maxLines: 1,
            // 是否自动更正
            autocorrect: true,
            // 是否自动对焦
            autofocus: true,
            // 是否是密码
            obscureText: false,
            // 文本对齐方式
            textAlign: TextAlign.center,
            // 输入文本的样式
            style: TextStyle(fontSize: 26.0, color: Colors.green),
            // 文本内容改变时回调
            onChanged: (text) {
              print('文本内容改变时回调 $text');
            },
            // 内容提交时回调
            onSubmitted: (text) {
              print('内容提交时回调 $text');
            },
            enabled: true, // 是否禁用
            decoration: InputDecoration(// 添加装饰效果
                fillColor: Colors.grey.shade200,// 添加灰色填充色
                filled: true,
                helperText: '用户名',
                prefixIcon: Icon(Icons.person),
                // 左侧图标
                suffixText: '用户名'),
                // 右侧文本提示
          ),
        ),
      ),
    ),
  );
 }
}
```

上述示例代码的视图展现大致如图 5-19 所示。

5.3.2 Card（卡片组件）

Card 即卡片组件块，内容可以由大多数类型的 Widget 构成，但通常与 ListTile 一起使用。Card 有一个 child，但它可以是支持多个 child 的列、行、列表、网格或其他小部件。默认情况下，Card 将其大小缩小为 0 像

图 5-19 TextField 组件应用示例

素。你可以使用 SizedBox 来限制 Card 的大小。在 Flutter 中，Card 具有圆角和阴影，这让它看起来有立体感。Card 组件属性见表 5-14。

表 5-14 Card 组件属性及描述

属性名	类型	默认值	说明
child	Widget		组件的子元素，任意 Widget 都可以
margin	EdgeInsetsGeometry		围绕在 decoration 和 child 之外的空白区域，不属于内容区域
shape	ShapeBorder	RoundedRectangleBorder	Card 的阴影效果，默认的阴影效果为圆角的长方形边。弧度为 4.0

示例代码如下：

```
import 'package:flutter/material.dart';

void main() {
  runApp(new MaterialApp(
    title: 'Card 布局示例 ',
    home: new MyApp(),
  ));
}

class MyApp extends StatelessWidget {
  @override
  Widget build(BuildContext context) {

    var card = new SizedBox(
      height: 250.0,
      child: new Card(
        child: new Column(
          children: <Widget>[
            new ListTile(
              title: new Text(
                ' 深圳市南山区深南大道 35 号 ',style: new TextStyle(fontWeight:
                  FontWeight.w300),
              ),
              subtitle: new Text(' 创想科技有限公司 '),
              leading: new Icon(
                Icons.home,
                color: Colors.lightBlue,
              ),
            ),
            new Divider(),
            new ListTile(
              title: new Text(
                ' 深圳市罗湖区沿海大道 32 号 ',style: new TextStyle(fontWeight:
                  FontWeight.w300),
              ),
              subtitle: new Text(' 一木培训机构 '),
```

```
          leading: new Icon(
            Icons.school,
            color: Colors.lightBlue,
          ),
        ),
        new Divider(),
      ],
    ),
  ),
);


  return new Scaffold(
    appBar: new AppBar(
      title: new Text('Card布局示例'),
    ),
    body: new Center(
      child: card,
    ),
  );
 }
}
```

上述示例代码的视图展现大致如图 5-20 所示。

图 5-20　Card 组件应用示例

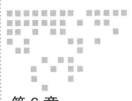

Chapter 6 第 6 章

Cupertino 风格组件

Cupertino 风格组件即 iOS 风格组件,主要有 CupertinoTabBar、CupertinoPageScaffold、CupertinoTabScaffold、CupertinoTabView 等。目前组件库还没有 Material Design 风格组件丰富。但作为一种补充,可以根据实际场景选择使用即可。

本章所涉及的组件及内容有:

❏ CupertinoActivityIndicator 组件
❏ CupertinoAlertDialog 对话框组件
❏ CupertinoButton 按钮组件
❏ Cupertino 导航组件集 (CupertinoTabBar、CupertinoPageScaffold、CupertinoTabScaffold、CupertinoTabView 等)

6.1 CupertinoActivityIndicator 组件

CupertinoActivityIndicator 是一个 iOS 风格的 loading 指示器,通常用来做加载等待的效果展示。

CupertinoActivityIndicator 组件属性如下所示:

属性名	类型	默认值	说明
radius	double	10.0	加载图形的半径值,值越大图形越大。
animating	bool	true	是否播放加载动画,通常用来做动画控制处理。比如:需要加载数据时置为 true,数据加载完成置为 false。

使用 Cupertino 组件需要导入 Cupertino 库,代码如下所示:

```
import 'package:flutter/cupertino.dart';
```

示例代码如下：

```
import 'package:flutter/material.dart';
import 'package:flutter/cupertino.dart';

void main() => runApp(MyApp());

class MyApp extends StatelessWidget {
  @override
  Widget build(BuildContext context) {
    return MaterialApp(
      title: 'CupertinoActivityIndicator 示例',
      home: Scaffold(
        appBar: AppBar(
          title: Text('CupertinoActivityIndicator 示例'),
        ),
        body: Center(
          child: CupertinoActivityIndicator(
            radius: 60.0,// 值越大加载的图形越大
          ),
        ),
      ),
    );
  }
}
```

上述示例代码的视图展现大致如图 6-1 所示。

6.2　CupertinoAlertDialog 对话框组件

CupertinoAlertDialog 和 Material Design 风格里的 AlertDialog 对话框相似，只是样式不同。共同点是不仅仅有提示内容，还有一些操作按钮，如确定和取消等。内容部分可以用 SingleChildScrollView 进行包裹。操作按钮建议用 CupertinoDialogAction 组件，这样显示更协调一些。

图 6-1　CupertinoActivityIndicator 组件应用示例

组件的属性及描述如下所示：

属性名	类型	说明
actions	List<Widget>	对话框底部操作按钮。例如确定、取消。
title	Widget	通常是一个文本组件。
content	Widget	内容部分，通常为对话框的提示内容。

编写一个 iOS 风格的删除确认的示例，完整的示例代码如下：

```dart
import 'package:flutter/material.dart';
import 'package:flutter/cupertino.dart';

void main() => runApp(MyApp());

class MyApp extends StatelessWidget {
  @override
  Widget build(BuildContext context) {
    return MaterialApp(
      title: 'CupertinoAlertDialog组件示例',
      home: Scaffold(
        appBar: AppBar(
          title: Text('CupertinoAlertDialog组件示例'),
        ),
        body: Center(
          child: CupertinoAlertDialog(
            title: Text('提示'), // 对话框标题
            content: SingleChildScrollView(
              // 对话框内容部分
              child: ListBody(
                children: <Widget>[
                  Text('是否要删除？'),
                  Text('一旦删除数据不可恢复！'),
                ],
              ),
            ),
            actions: <Widget>[
              CupertinoDialogAction(
                child: Text('确定'),
                onPressed: () {},
              ),
              CupertinoDialogAction(
                child: Text('取消'),
                onPressed: () {},
              ),
            ],
          ),
        ),
      ),
    );
  }
}
```

上述示例代码的视图展现大致如图6-2所示。

图6-2 CupertinoAlertDialog组件应用示例

6.3 CupertinoButton 按钮组件

CupertinoButton 展示 iOS 风格的按钮。它可以响应按下事件，并且按下时会带一个触

摸的效果。常用属性如下所示：

属性名	类型	默认值	说明
color	Color		组件的颜色。
disabledColor	Color	ThemeData.disabledColor	组件的禁用颜色，默认为主题里的禁用颜色。
onPressed	VoidCallback	null	当按钮按下时会触发此回调事件。
child	Widget		按钮的 Child 通常为一个 Text 文本组件，用来显示按钮的文本。
enable	bool	true	按钮是否为禁用状态。

示例代码如下：

```
import 'package:flutter/material.dart';
import 'package:flutter/cupertino.dart';

void main() => runApp(MyApp());

class MyApp extends StatelessWidget {
  @override
  Widget build(BuildContext context) {
    return MaterialApp(
      title: 'CupertinoButton 组件示例',
      home: Scaffold(
        appBar: AppBar(
          title: Text('CupertinoButton 组件示例'),
        ),
        body: Center(
          child: CupertinoButton(
            child: Text(// 按钮 label
              'CupertinoButton',
            ),
            color: Colors.blue,// 按钮颜色
            onPressed: (){},// 按下事件回调
          ),
        ),
      ),
    );
  }
}
```

上述示例代码的视图展现大致如图 6-3 所示，当按下图中按钮后，会触发事件回调 onPressed 方法。

图 6-3　CupertinoButton 组件应用示例

6.4　Cupertino 导航组件集

Cupertino 导航相关的组件也非常丰富。由于它们之间有关联关系，所以本节把导航相

关的组件放在一起说明。下面分别说明这些组件的属性及描述。

1. CupertinoTabScaffold

选项卡组件，将选项卡按钮及选项卡视图绑定在一起。常用属性如下所示：

属性名	类型	说明
tabBar	CupertinoTabBar	选项卡按钮，通常由图标加文本组成。
tabBuilder	IndexedWidgetBuilder	选项卡视图构造器。

2. CupertinoTabBar

选项卡按钮，通常由 BottomNavigationBarItem 组成包含图标加文本，效果如图 6-4 所示。常用属性如下所示：

属性名	类型	说明
items	List<BottomNavigationBarItem>	选项卡按钮数据集合，通常由图标加文本组成。
backgroundColor	Color	选项卡按钮背景色。
activeColor	Color	选中的选项卡按钮前景色。
iconSize	double	选项卡图标大小。

图 6-4　CupertinoTabBar 组件效果图

3. CupertinoTabView

选项卡视图，常用属性如下所示：

属性名	类型	说明
builder	WidgetBuilder	选项卡视图构造器。
routes	Map<String，WidgetBuilder>	选项卡视图路由。

4. CupertinoPageScaffold

页面的基本布局结构，包含内容和导航栏。常用属性如下所示：

属性名	类型	说明
backgroundColor	Color	页面背景色。
navigationBar	ObstructingPreferredSizeWidget	顶部导航栏按钮，包含左中右三个子组件，比如：页面中的返回按钮。
child	Widget	页面的主要内容。

5. CupertinoNavigationBar

导航栏结构组件，效果图如图 6-5 所示。常用属性如下所示：

属性名	类型	说明
middle	Widget	导航栏中间组件，通常为页面标题。
trailing	Widget	导航栏右边组件，通常为菜单按钮。
leading	Widget	导航栏左边组件，通常为返回按钮。

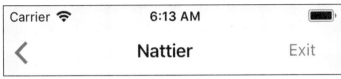

图 6-5　CupertinoNavigationBar 组件效果图

6. 导航综合示例

接下来编写一个使用导航组件的综合例子。步骤如下：

步骤 1：添加最外层导航选项卡，使用 CupertinoTabScaffold 组件。它有两个属性 tabBar 和 tabBuilder，分别用来绑定底部选项卡及所对应的视图。

步骤 2：添加底部选项卡按钮，使用 CupertinoTabBar 组件。需要指定选项卡项，每个项包含一个图标及一个文本。代码结构如下所示：

```
tabBar: CupertinoTabBar(
  items: [
    BottomNavigationBarItem(
      // 指定图标
      // 指定文本
    ),
    // 根据需要添加更多按钮
  ],
),
```

此部分代码效果图如图 6-6 所示。

图 6-6　导航栏底部效果图

步骤 3：每个选项卡需要绑定一个视图，使用 CupertinoTabView 组件。这需要使用分支语句来判断当前选中的是哪一个选项卡。代码如下所示：

```
tabBuilder: (context, index) {
  // 选项卡绑定的视图
  return CupertinoTabView(
    builder: (context) {
      switch (index) {
```

```
          case 0:
            return HomePage();
            break;
          case 1:
            return ChatPage();
            break;
          default:
            return Container();
        }
      },
    );
  },
```

步骤4：编写第一个选项卡页面即主页，将其命名为HomePage。页面需要一个CupertinoPageScaffold组件来包裹，指定页面顶部导航内容及内容区域。代码如下所示：

```
class HomePage extends StatelessWidget {
  @override
  Widget build(BuildContext context) {
    return CupertinoPageScaffold(
      navigationBar: CupertinoNavigationBar(
        // 只加标题
      ),
      child: Center(
        // 内容区域
      ),
    );
  }
}
```

此部分代码效果图如图6-7所示。

步骤5：编写第二个选项卡页面即聊天页面，将其命名为ChatPage。页面需要一个CupertinoPageScaffold组件来包裹，指定页面顶部导航内容（包含左中右三块），及内容区域。代码结构如下所示：

```
// 聊天页面
class ChatPage extends StatelessWidget {
  @override
  Widget build(BuildContext context) {
    return CupertinoPageScaffold(
      navigationBar: CupertinoNavigationBar(
        // 导航栏 包含左中右三部分
      ),
      child: Center(
        // 内容区域
      ),
    );
  }
}
```

此部分代码的效果图如图 6-8 所示。

图 6-7　主页效果图

图 6-8　聊天页面效果图

步骤 6：组装所有内容，完整的代码如下所示：

```
import 'package:flutter/material.dart';
import 'package:flutter/cupertino.dart';

void main() => runApp(MyApp());

class MyApp extends StatelessWidget {
  @override
  Widget build(BuildContext context) {
    return MaterialApp(
      title: 'Cupertino导航组件集',
      theme: ThemeData.light(), // 浅色主题
      home: MyPage(),
    );
  }
}

class MyPage extends StatefulWidget {
  @override
  _MyPageState createState() => _MyPageState();
}

class _MyPageState extends State<MyPage> {
  @override
```

```
Widget build(BuildContext context) {
  // 最外层导航选项卡
  return CupertinoTabScaffold(
    // 底部选项卡
    tabBar: CupertinoTabBar(
      backgroundColor: CupertinoColors.lightBackgroundGray, // 选项卡背景色
      items: [
        // 选项卡项 包含图标及文字
        BottomNavigationBarItem(
          icon: Icon(CupertinoIcons.home),
          title: Text('主页'),
        ),
        BottomNavigationBarItem(
          icon: Icon(CupertinoIcons.conversation_bubble),
          title: Text('聊天'),
        ),
      ],
    ),
    tabBuilder: (context, index) {
      // 选项卡绑定的视图
      return CupertinoTabView(
        builder: (context) {
          switch (index) {
            case 0:
              return HomePage();
              break;
            case 1:
              return ChatPage();
              break;
            default:
              return Container();
          }
        },
      );
    },
  );
}

// 主页
class HomePage extends StatelessWidget {
  @override
  Widget build(BuildContext context) {
    return CupertinoPageScaffold(
      // 基本布局结构，包含内容和导航栏
      navigationBar: CupertinoNavigationBar(
        // 导航栏 只包含中部标题部分
        middle: Text("主页"),
      ),
      child: Center(
        child: Text(
          '主页',
          style: Theme.of(context).textTheme.button,
        ),
      ),
```

```
    );
  }
}

// 聊天页面
class ChatPage extends StatelessWidget {
  @override
  Widget build(BuildContext context) {
    return CupertinoPageScaffold(
      navigationBar: CupertinoNavigationBar(
        // 导航栏 包含左中右三部分
        middle: Text(" 聊天面板 "),// 中间标题
        trailing: Icon(CupertinoIcons.add),// 右侧按钮
        leading: Icon(CupertinoIcons.back),// 左侧按钮
      ),
      child: Center(
        child: Text(
          '聊天面板',
          style: Theme.of(context).textTheme.button,
        ),
      ),
    );
  }
}
```

上述示例的效果如图 6-9 所示。

图 6-9　Cupertino 导航组件集示例效果图

Chapter 7 第 7 章

页面布局

前面几章介绍了基础组件及各种主题风格组件，这样再加上一些布局元素，就可以实现一些基础的页面了。本章主要讲解布局及装饰组件的基本用法。最后再配合一个综合布局示例，来完整地演示如何编写一个复杂的页面。

主要的布局及装饰组件参见表 7-1，本章将按照以下分类介绍这些组件：

❏ 基础布局处理

❏ 宽高尺寸处理

❏ 列表及表格布局

❏ 其他布局处理

❏ 布局综合示例

表 7-1 布局及装饰组件的说明

组件名称	中文名称	简单说明
Align	对齐布局	指定 child 的对齐方式
AspectRatio	调整宽高比	根据设置的宽高比调整 child
Baseline	基准线布局	所有 child 底部所在的同一条水平线
Center	居中布局	child 处于水平和垂直方向的中间位置
Column	垂直布局	对 child 在垂直方向进行排列
ConstrainedBox	限定宽高	限定 child 的最大值
Container	容器布局	容器布局是一个组合的 Widget，包含定位和尺寸
FittedBox	缩放布局	缩放以及位置调整
FractionallySizedBox	百分比布局	根据现有空间按照百分比调整 child 的尺寸
GridView	网格布局	对多行多列同时进行操作

(续)

组件名称	中文名称	简单说明
IndexedStack	栈索引布局	IndexedStack 继承自 Stack，显示第 index 个 child，其他 child 都是不可见的
LimitedBox	限定宽高布局	对最大宽高进行限制
ListView	列表布局	用列表方式进行布局，比如多项数据的场景
Offstage	开关布局	控制是否显示组件
OverflowBox	溢出父容器显示	允许 child 超出父容器的范围显示
Padding	填充布局	处理容器与其 child 之间的间距
Row	水平布局	对 child 在水平方向进行排列
SizedBox	设置具体尺寸	用一个特定大小的盒子来限定 child 宽度和高度
Stack/Alignment	Alignment 栈布局	根据 Alignment 组件的属性将 child 定位在 Stack 组件上
Stack/Positioned	Positioned 栈布局	根据 Positioned 组件的属性将 child 定位在 Stack 组件上
Table	表格布局	使用表格的行和列进行布局
Transform	矩阵转换	做矩阵变换，对 child 做平移、旋转、缩放等操作
Wrap	按宽高自动换行	按宽度或者高度，让 child 自动换行布局

7.1 基础布局处理

基础布局组件包括容器布局，各种缩放、排列方式的组件，下面分别详述。

7.1.1 Container（容器布局）

Container（容器布局）在 Flutter 里大量使用，它是一个组合 Widget，内部有绘制 Widget、定位 Widget 和尺寸 Widget。Container 组件的常见属性请参考 4.1 节"容器组件"的介绍。接下来我们看一个用 Container 进行容器布局的综合小例子。具体实现步骤如下。

步骤 1：在 helloworld 工程下新建一个 images 文件夹，并放入三张图片，如图 7-1 所示。

步骤 2：打开工程根目录下的工程配置文件，如图 7-2 所示。

步骤 3：修改工程配置文件，添加图片的路径配置，如图 7-3 所示。

图 7-1 添加三张图片

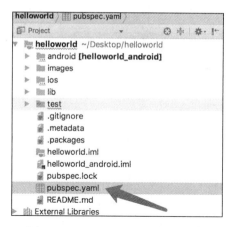

图 7-2 pubspec.yaml 工程配置文件

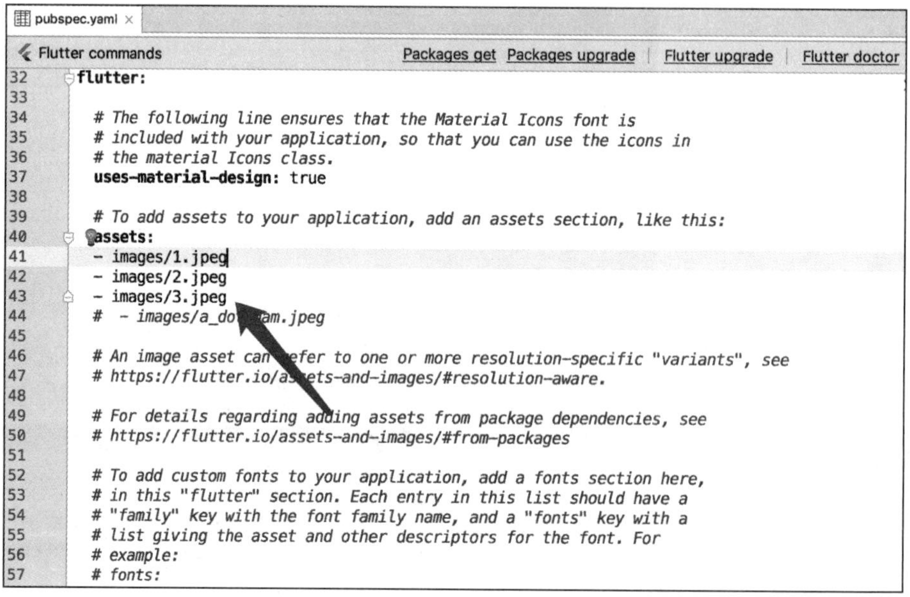

图 7-3 修改图片资源配置

> **注意** 图片资源配置项不要多加空格，否则运行会报错。另外路径一定要配置正确，并且要确保在指定路径下有此图片，否则运行同样会报错。

步骤 4：编写 Container 容器布局示例代码，完整的示例代码如下：

```
import 'package:flutter/material.dart';

void main() => runApp(
    new MaterialApp(
        title: 'Container 布局容器示例',
```

```
      home: new LayoutDemo(),
    ),
  );

class LayoutDemo extends StatelessWidget {
  @override
  Widget build(BuildContext context) {
    //返回一个Container对象
    Widget container = new Container(
      //添加装饰效果
      decoration: new BoxDecoration(
        color: Colors.grey,
      ),
      //子元素指定为一个垂直水平嵌套布局的组件
      child: new Column(
        children: <Widget>[
          new Row(
            children: <Widget>[
              //使用Expanded防止内容溢出
              new Expanded(
                child: new Container(
                  width: 150.0,
                  height: 150.0,
                  //添加边框样式
                  decoration: new BoxDecoration(
                    //上下左右边框设置为宽度10.0,颜色为蓝灰色
                    border: new Border.all(width: 10.0, color: Colors.blueGrey),
                    //上下左右边框弧度设置为8.0
                    borderRadius:
                        const BorderRadius.all(const Radius.circular(8.0)),
                  ),
                  //上下左右增加边距
                  margin: const EdgeInsets.all(4.0),
                  //添加图片
                  child: new Image.asset('images/1.jpeg'),
                ),
              ),
              new Expanded(
                child: new Container(
                  width: 150.0,
                  height: 150.0,
                  decoration: new BoxDecoration(
                    border: new Border.all(width: 10.0, color: Colors.blueGrey),
                    borderRadius:
                        const BorderRadius.all(const Radius.circular(8.0)),
                  ),
                  margin: const EdgeInsets.all(4.0),
                  child: new Image.asset('images/2.jpeg'),
                ),
              ),
```

```
          ],
        ),
        new Row(
          children: <Widget>[
            new Expanded(
              child: new Container(
                width: 150.0,
                height: 150.0,
                decoration: new BoxDecoration(
                  border: new Border.all(width: 10.0, color: Colors.blueGrey),
                  borderRadius:
                      const BorderRadius.all(const Radius.circular(8.0)),
                ),
                margin: const EdgeInsets.all(4.0),
                child: new Image.asset('images/3.jpeg'),
              ),
            ),
            new Expanded(
              child: new Container(
                width: 150.0,
                height: 150.0,
                decoration: new BoxDecoration(
                  border: new Border.all(width: 10.0, color: Colors.blueGrey),
                  borderRadius:
                      const BorderRadius.all(const Radius.circular(8.0)),
                ),
                margin: const EdgeInsets.all (4.0),
                child: new Image.asset ('images/ 2.jpeg'),
              ),
            ),
          ],
        ),
      ],
    ),
  );

  return new Scaffold(
    appBar: new AppBar(
      title: new Text('Container布局容器示例'),
    ),
    body: container,
  );
 }
}
```

上述示例代码的视图展现大致如图 7-4 所示。

图 7-4　Container 容器布局示例

7.1.2　Center（居中布局）

在 Center 居中布局中，子元素处于水平和垂直方向的中间位置。

示例代码如下：

```
import 'package:flutter/material.dart';

void main() => runApp(
      new MaterialApp(
        title: 'Center居中布局示例',
        home: new LayoutDemo(),
      ),
    );

class LayoutDemo extends StatelessWidget {
  @override
  Widget build(BuildContext context) {
    return new Scaffold(
      appBar: new AppBar(
        title: new Text('Center居中布局示例'),
      ),
      body: new Center(
        child: new Text(
```

```
          'Hello Flutter',
          style: TextStyle(
            fontSize: 36.0,
          ),
        ),
      ),
    );
  }
}
```

> **注意** Center 属性在通常情况下可直接添加子元素，不需要做额外属性设置。

上述示例代码的视图展现大致如图 7-5 所示。

图 7-5 Center 居中布局示例

7.1.3 Padding（填充布局）

Padding 即为填充组件，用于处理容器与其子元素之间的间距，与 padding 属性对应的是 margin 属性，margin 处理容器与其他组件之间的间距。Padding 组件的常见属性如下所示：

属性名	类型	说明
padding	EdgeInsetsGeometry	填充值可以使用 EdgeInsets 方法，例如：EdgeInsets.all(6.0) 将容器上下左右填充设置为 6.0，也可以使用 EdgeInsets.only 方法来单独设置某一边的间距。

接下来我们写一个例子，容器里嵌套了一个容器，两个容器分别加了一个边框，以便测试 Padding 值。

示例代码如下：

```dart
import 'package:flutter/material.dart';

void main() => runApp(
      new MaterialApp(
        title: 'Padding 填充布局示例 ',
        home: new LayoutDemo(),
      ),
    );

class LayoutDemo extends StatelessWidget {
  @override
  Widget build(BuildContext context) {
    return new Scaffold(
      appBar: new AppBar(
        title: new Text('Padding 填充布局示例 '),
      ),
      body: new Center(

    child: new Container(
    width: 300.0,
      height: 300.0,
      padding: EdgeInsets.all(60.0),// 容器上下左右填充设置为 60.0
      decoration: new BoxDecoration(
        color: Colors.white,
        border: new Border.all(
          color: Colors.green,
          width: 8.0,
        ),
      ),
      child: new Container(
        width: 200.0,
        height: 200.0,
        decoration: new BoxDecoration(
          color: Colors.white,
          border: new Border.all(
            color: Colors.blue,
            width: 8.0,
          ),
        ),
        child: new FlutterLogo(),
      ),
    ),
    ),
    );
  }
```

}

上述示例代码的 Padding 值为 60.0。视图展现大致如图 7-6 所示。

接着再将 Padding 值改为 6.0。注意此时的值小了很多，我们会发现两个容器之间的间距小了很多，如图 7-7 所示。

图 7-6　Padding 填充布局示例图一

图 7-7　Padding 填充布局示例图二

7.1.4　Align（对齐布局）

Align 组件即对齐组件，能将子组件按指定方式对齐，并根据子组件的大小调整自己的大小。Align 组件的属性及说明如下所示：

属性名	值	说明
bottomCenter	(0.5, 1.0)	底部中心
bottomLeft	(0.0, 1.0)	左下角
bottomRight	(1.0, 1.0)	右下角
center	(0.5, 0.5)	水平垂直居中
centerLeft	(0.0, 0.5)	左边缘中心
centerRight	(1.0, 0.5)	右边缘中心
topCenter	(0.5, 0.0)	顶部中心
topLeft	(0.0, 0.0)	左上角
topRight	(1.0, 0.0)	右上角

这里我们写一个例子，放入多张图片，这些图片分别放置在不同的位置。具体步骤如下。
步骤 1：在 helloworld 工程下新建一个 images 文件夹，并放入 9 张图片，如图 7-8 所示。
步骤 2：打开工程根目录下的文件。如图 7-9 所示。

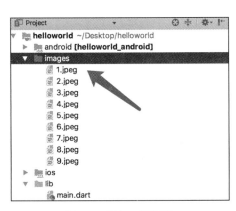

图 7-8　添加 9 张图片

图 7-9　打开 pubspec.yaml 工程配置文件

步骤 3：修改工程配置文件，添加图片的路径配置。如图 7-10 所示。

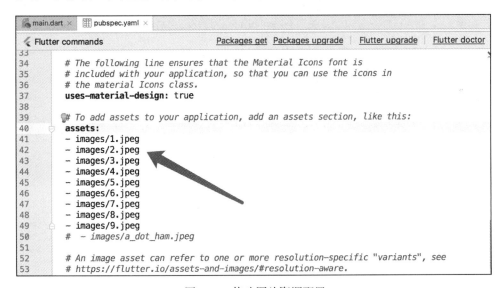

图 7-10　修改图片资源配置

> **注意**　图片资源配置项不要多加空格，否则运行报错。另外路径一定要配置正确，并且要确保在指定路径下有此图片，否则运行同样报错。

接着添加一个 Stack 组件，暂时不用管 Stack 组件是用来做什么的，我们需要在这个组件

里放几张图片,分别放在左上角、右上角、左下角、右下角和中间。完整的示例代码如下:

```dart
import 'package:flutter/material.dart';
class LayoutDemo extends StatelessWidget {
  @override
  Widget build(BuildContext context) {
    return new Scaffold(
      appBar: new AppBar(
        title: new Text('Align 对齐布局示例'),
      ),
      body: new Stack(
        children: <Widget>[
          //左上角
          new Align(
            alignment: new FractionalOffset(0.0, 0.0),
            child: new Image.asset('images/1.jpeg',width: 128.0,height: 128.0,),
          ),
          //右上角
          new Align(
            alignment: FractionalOffset(1.0,0.0),
            child: new Image.asset('images/1.jpeg',width: 128.0,height: 128.0,),
          ),
          // 水平垂直方向居中
          new Align(
            alignment: FractionalOffset.center,
            child: new Image.asset('images/3.jpeg',width: 128.0,height: 128.0,),
          ),
          //左下角
          new Align(
            alignment: FractionalOffset.bottomLeft,
            child: new Image.asset('images/2.jpeg',width: 128.0,height: 128.0,),
          ),
          //右下角
          new Align(
            alignment: FractionalOffset.bottomRight,
            child: new Image.asset('images/2.jpeg',width: 128.0,height: 128.0,),
          ),
        ]
      ),
    );
  }
}
void main() {
  runApp(
    new MaterialApp(
      title: 'Align 对齐布局示例',
      home: new LayoutDemo(),
    ),
  );
}
```

视图的展现大致如图 7-11 所示。

图 7-11　Align 对齐布局示例

7.1.5　Row（水平布局）

水平布局是一种常用的布局方式，我们主要使用 Row 组件来完成子组件在水平方向的排列。Row 组件常见属性如下所示：

属性名	类型	说明
mainAxisAlignment	MainAxisAlignment	主轴的排列方式。
crossAxisAlignment	CrossAxisAlignment	次轴的排列方式。
mainAxisSize	MainAxisSize	主轴应该占据多少空间。取值 max 为最大，min 为最小。
children	List<Widget>	组件子元素，它的本质是一个 List 列表。

对 Row 来说，水平方向是主轴，垂直方向是次轴，可以完全参照 Web 中的 Flex 布局，如图 7-12 所示。

示例代码如下：

```
import 'package:flutter/material.dart';

void main() => runApp(
  new MaterialApp(
    title: '水平布局示例',
    home: new LayoutDemo(),
  ),
);
```

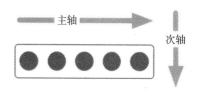

图 7-12　Row 组件的轴示意图

```
class LayoutDemo extends StatelessWidget {

  @override
  Widget build(BuildContext context) {

    return new Scaffold(
      appBar: new AppBar(
        title: new Text('水平布局示例'),
      ),
      body: new Row(
        children: <Widget>[
          new Expanded(
            child: new Text('左侧文本', textAlign: TextAlign.center),
          ),
          new Expanded(
            child: new Text('中间文本', textAlign:
              TextAlign.center),
          ),
          new Expanded(
            child: new FittedBox(
              fit: BoxFit.contain,
              child: const FlutterLogo(),
            ),
          ),
        ],
      ),
    );

  }
}
```

上述示例代码的视图展现大致如图 7-13 所示。

7.1.6 Column（垂直布局）

垂直布局是一种常用的布局方式，我们主要使用 Column 组件来完成对子组件纵向的排列。Column 组件常见属性如下所示：

图 7-13　水平布局示例

属性名	类型	说明
mainAxisAlignment	MainAxisAlignment	主轴的排列方式。
crossAxisAlignment	CrossAxisAlignment	次轴的排列方式。
mainAxisSize	MainAxisSize	主轴应该占据多少空间。取值 max 为最大，min 为最小。
children	List<Widget>	组件子元素，它的本质是一个 List 列表。

对于 Column 来说，垂直方向就是主轴，水平方向就是次轴，可以完全参照 Web 中的 Flex 布局，如图 7-14 所示。

第一个垂直布局的示例代码如下：

```
import 'package:flutter/material.dart';

void main() => runApp(
  new MaterialApp(
    title: '水平布局示例',
    home: new LayoutDemo(),
  ),
);

class LayoutDemo extends StatelessWidget {

  @override
  Widget build(BuildContext context) {

    return new Scaffold(
      appBar: new AppBar(
        title: new Text('垂直布局示例一'),
      ),
      body: new Column(
        children: <Widget>[
          new Text('Flutter'),
          new Text('垂直布局'),
          new Expanded(
            child: new FittedBox(
              fit: BoxFit.contain,
              child: const FlutterLogo(),
            ),
          ),
        ],
      ),
    );
  }
}
```

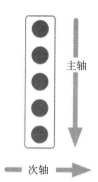

图 7-14　Column 组件的轴示意图

上述示例代码的视图展现大致如图 7-15 所示。

若要增加水平方向靠左对齐，主轴方向最小化处理，其中最后一行文字字号专门放大，可以将第一个示例中 Column 里的内容换成如下代码：

```
crossAxisAlignment: CrossAxisAlignment.start,// 水平方向靠左对齐
mainAxisSize: MainAxisSize.min,// 主轴方向最小化处理
children: <Widget>[
  new Text('Flutter是谷歌的移动UI框架'),
  new Text('可以快速在iOS和Android上构建高质量的原生用户界面'),
  new Text('Flutter可以与现有的代码一起工作。在全世界'),
  new Text('Flutter正在被越来越多的开发者和组织使用'),
  new Text('并且Flutter是完全免费、开源的。'),
  new Text('Flutter!', style: TextStyle(fontSize: 36.0,)),// 放大字号
```

第二个示例代码的视图展现大致如图 7-16 所示。文字全部向左对齐，上面几行文字都

偏小，只有最后一行专门放大处理后文字显得比较大。

图 7-15　垂直布局示例一

图 7-16　垂直布局示例二

7.1.7　FittedBox（缩放布局）

FittedBox 组件主要做两件事情，缩放（Scale）和位置调整（Position）。

FittedBox 会在自己的尺寸范围内缩放并调整 child 位置，使 child 适合其尺寸。做过移动端的读者可能会联想到 ImageView 组件，它是将图片在其范围内按照规则进行缩放位置调整。FittedBox 跟 ImageView 有些类似，可以猜出，它肯定有一个类似于 ScaleType 的属性。

布局行为分两种情况：

❏ 如果外部有约束的话，按照外部约束调整自身尺寸，然后缩放调整 child，按照指定的条件进行布局。

❏ 如果没有外部约束条件，则跟 child 尺寸一致，指定的缩放以及位置属性将不起作用。

有 fit 和 alignment 两个重要属性，如下所示。

fit：缩放的方式，默认的属性是 BoxFit.contain，child 在 FittedBox 范围内，尽可能大，但是不超出其尺寸。这里需要注意一点，contain 是在保持着 child 宽高比的大前提下，尽可能填满。一般情况下，宽度或者高度达到最大值时，就会停止缩放。Fit 的各个属性对应的布局示意图如图 7-17 所示。

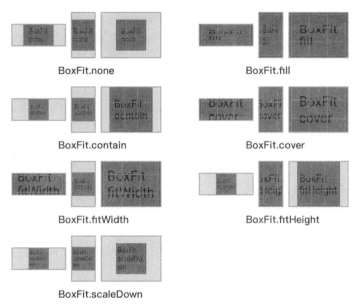

图 7-17　fit 属性各个取值示意图

BoxFit.none 没有任何填充模式，如图 7-18 所示。

BoxFit.fill 不按宽高比填充模式，内容不会超过容器范围，效果如图 7-19 所示。

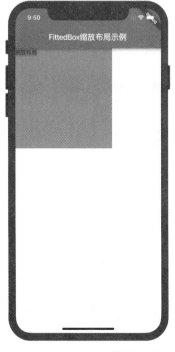 

图 7-18　BoxFit.none 填充效果图　　　　图 7-19　BoxFit.fill 填充效果图

BoxFit.contain 按宽高比等比填充模式，内容不会超过容器范围，效果如图 7-20 所示。

BoxFit.cover 按原始尺寸填充整个容器模式，内容有可能会超过容器范围，效果如图 7-21 所示。

图 7-20　BoxFit.contain 填充效果图　　图 7-21　BoxFit.cover 填充效果图

BoxFit.width 及 BoxFit.height 分别是按宽 / 高填充整个容器的模式，内容不会超过容器范围，效果如图 7-22 所示。

图 7-22　BoxFit.width 及 BoxFit.height 填充效果图

图 7-22 左侧为按宽度填充，右侧为按高度填充

BoxFit.scaleDown 会根据情况缩小范围，有时和 BoxFit.contain 一样，有时和 BoxFit.none 一样，内容不会超过容器范围，如图 7-23 所示。

alignment：设置对齐方式，默认的属性是 Alignment.center，居中显示 child。填充布局的示例代码如下：

```
import 'package:flutter/material.dart';
class LayoutDemo extends StatelessWidget {
  @override
  Widget build(BuildContext context) {
    return new Scaffold(
      appBar: new AppBar(
        title: new Text('FittedBox 缩放布局示例'),
      ),
      body: new Container(
        color: Colors.grey,
        width: 250.0,
        height: 250.0,
        child: new FittedBox(
          fit: BoxFit.contain,// 改变填充属性值会得到
                              不同的效果
          alignment: Alignment.topLeft,
          child: new Container(
            color: Colors.deepOrange,
            child: new Text(" 缩放布局 "),
          ),
        ),
      ),
    );
  }
}
void main() {
  runApp(
    new MaterialApp(
      title: 'FittedBox 缩放布局示例',
      home: new LayoutDemo(),
    ),
  );
}
```

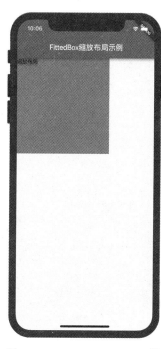

图 7-23 BoxFit.scaleDown 填充效果图

7.1.8 Stack/Alignment

Stack 组件的每一个子组件要么定位，要么不定位，定位的子组件是用 Positioned 组件包裹的。Stack 组件本身包含所有不定位的子组件，子组件根据 alignment 属性定位（默认为左上角）。然后根据定位的子组件的 top、right、bottom 和 left 属性将它们放置在 Stack 组件上。Stack 组件的主要属性如下所示。

属性名	类型	默认值	说明
alignment	AlignmentGeometry	Alignment.topLeft	定位位置有以下几种： ● bottomCenter 底部中间位置。 ● bottomLeft 底部左侧位置。 ● bottomRight，底部右侧位置。 ● center，正中间位置。 ● centerLeft，中间居左位置。 ● centerRight，中间居右位置。 ● topCenter，上部居中位置。 ● topLeft，上部居左位置。 ● topRight，上部居右位置。

采用 Alignment 方式布局示例代码如下：

```
import 'package:flutter/material.dart';

void main() {
  runApp(new MaterialApp(
    title: 'Stack 布局示例 Alignment',
    home: new MyApp(),
  ));
}

class MyApp extends StatelessWidget {
  @override
  Widget build(BuildContext context) {

    var stack = new Stack(
      // 子组件左上角对齐
      alignment: Alignment.topLeft,
      children: <Widget>[
        // 底部添加一个头像
        new CircleAvatar(
          backgroundImage: new AssetImage('images/1.jpeg'),
          radius: 100.0,
        ),
        // 上面加一个容器，容器里再放一段文本
        new Container(
          decoration: new BoxDecoration(
            color: Colors.black38,
          ),
          child: new Text(
            '我是超级飞侠',
            style: new TextStyle(
              fontSize: 22.0,
              fontWeight: FontWeight.bold,
              color: Colors.white,
            ),
          ),
        ),
      ],
```

```
    );

    return new Scaffold(
      appBar: new AppBar(
        title: new Text('Stack 层叠布局示例'),
      ),
      body: new Center(
        child: stack,
      ),
    );
  }
}
```

上述示例代码的视图展现大致如图 7-24 所示。Stack 子组件是层叠的关系，所以示例里文本会在图片之上。

图 7-24　Stack 组件布局应用示例

> **注意**　示例中的图片可以使用之前章节使用的图片资源。

7.1.9　Stack/Positioned

Positioned 组件是用来定位的。Stack 使用 Positioned 布局主要是因为在 Stack 组件里面需要包裹一个定位组件，Positioned 组件属性如下所示：

属性名	类型	说明
top	double	子元素相对顶部边界距离。
bottom	double	子元素相对底部边界距离。
left	double	子元素相对左侧边界距离。
right	double	子元素相对右侧边界距离。

Positioned 定位布局的示例代码如下：

```dart
import 'package:flutter/material.dart';

void main() {
  runApp(new MaterialApp(
    title: '层叠定位布局示例',
    home: new MyApp(),
  ));
}

class MyApp extends StatelessWidget {
  @override
  Widget build(BuildContext context) {
    return new Scaffold(
      appBar: new AppBar(
        title: new Text('层叠定位布局示例'),
      ),
      body: new Center(
        //添加层叠布局
        child: new Stack(
          children: <Widget>[
            //添加网络图片
            new Image.network('https://timgsa.baidu.com/timg?image&quality=80&size=b9999_10000&sec=1541655494719&di=6b49d24b5172a34828b9d6506e4bf100&imgtype=0&src=http%3A%2F%2Fn.sinaimg.cn%2Fsinacn11%2F266%2Fw640h426%2F20180813%2Fce56-hhqtawx8254771.jpg'),
            //设置定位布局
            new Positioned(
              bottom: 50.0,
              right: 50.0,
              child: new Text(
                'hi flutter',
                style: new TextStyle(
                  fontSize: 36.0,
                  fontWeight: FontWeight.bold,
                  fontFamily: 'serif',
                  color: Colors.white,
                ),
              )
            ),
```

```
          ],
        ),
      ),
    );
  }
}
```

上述示例代码的视图展现大致如图 7-25 所示。

图 7-25　Stack 层叠定位布局示例

注意　示例中的图片，可以在网上任意复制一个图片地址即可。

7.1.10　IndexedStack

IndexedStack 继承自 Stack，它的作用是显示第 index 个 child，其他 child 都是不可见的，所以 IndexedStack 的尺寸永远是和最大的子节点尺寸一致。由于 IndexedStack 是继承自 Stack 的，所以它只比 Stack 多了一个 index 属性，即对应 child 的索引。

这里我们改造 Stack 布局示例，把 Stack 组件替换为 IndexedStack 组件，同时添加 index 属性值，当 index 设置为 1 时，运行结果只显示"我是超级飞侠"几个字及它的容器。因为 CircleAvatar 组件索引为 0，所以不显示。

示例的完整代码如下：

```dart
import 'package:flutter/material.dart';

void main() {
  runApp(new MaterialApp(
    title: 'IndexedStack布局示例',
    home: new MyApp(),
  ));
}

class MyApp extends StatelessWidget {
  @override
  Widget build(BuildContext context) {

    var stack = new IndexedStack(
      index: 1,// 设置索引为1就只显示文本内容了
      alignment: const FractionalOffset(0.2, 0.2),
      children: <Widget>[
        new CircleAvatar(
          backgroundImage: new AssetImage('images/1.jpeg'),
          radius: 100.0,
        ),
        new Container(
          decoration: new BoxDecoration(
            color: Colors.black38,
          ),
          child: new Text(
            '我是超级飞侠',
            style: new TextStyle(
              fontSize: 22.0,
              fontWeight: FontWeight.bold,
              color: Colors.white,
            ),
          ),
        ),
      ],
    );

    return new Scaffold(
      appBar: new AppBar(
        title: new Text('Stack层叠布局示例'),
      ),
      body: new Center(
        child: stack,
      ),
    );
  }
}
```

上述示例代码的视图展现大致如图7-26所示。

图 7-26　IndexedStack 组件布局应用示例

7.1.11　OverflowBox（溢出父容器显示）

通常，子组件是无法打破由父组件传递下来的约束条件，但是我们总会有一些情况下子组件的尺寸大于父组件。那怎么解决呢？在子组件及父组件之间再加入一个 OverflowBox 组件，那么子组件就不会以父组件为约束条件，而是以 OverflowBox 组件的属性为约束条件。OverflowBox 组件由其 minWidth、minHeight、maxWidth、maxHeight 四个属性给子组件划定了一个矩形的范围。子组件不能超出 OverflowBox 组件范围，但是可以溢出父组件范围显示。

OverflowBox 组件的主要属性如下所示：

属性名	类型	说明
alignment	AlignmentGeometry	对齐方式。
minWidth	double	允许 child 的最小宽度。
maxWidth	double	允许 child 的最大宽度。
minHeight	double	允许 child 的最小高度。
maxHeight	double	允许 child 的最大高度。

接下来，我们写一个例子，添加两个组件：绿色父组件和蓝灰色子组件，子组件会溢出父组件一定范围。如果没有 OverflowBox 组件，子组件蓝灰色的盒子是不会超出父组件

绿色的盒子的。完整的示例代码如下：

```
import 'package:flutter/material.dart';

class LayoutDemo extends StatelessWidget {
  @override
  Widget build(BuildContext context) {
    return new Scaffold(
      appBar: new AppBar(
        title: new Text('OverflowBox 溢出父容器显示示例'),
      ),
      body: Container(
        color: Colors.green,
        width: 200.0,
        height: 200.0,
        padding: const EdgeInsets.all(50.0),
        child: OverflowBox(
          alignment: Alignment.topLeft,
          maxWidth: 400.0,
          maxHeight: 400.0,
          child: Container(
            color: Colors.blueGrey,
            width: 300.0,
            height: 300.0,
          ),
        ),
      ));
  }
}
void main() {
  runApp(
    new MaterialApp(
      title: 'OverflowBox 溢出父容器显示示例 ',
      home: new LayoutDemo(),
    ),
  );
}
```

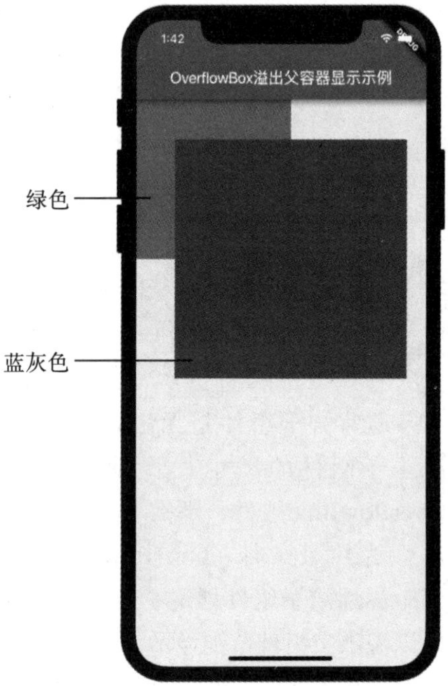

上述示例代码的视图展现大致如图 7-27 所示。结果显示出，子组件蓝灰色的盒子超出父组件绿色盒子了。

图 7-27　OverflowBox 组件布局示例

7.2　宽高尺寸处理

对组件具体尺寸的设置有多种方式，本节将详述这类组件。

7.2.1　SizedBox（设置具体尺寸）

SizedBox 组件是一个特定大小的盒子，这个组件强制它的 child 有特定的宽度和高度。

如果宽度或高度为 null，则此组件将调整自身大小以匹配该维度中 child 的大小。

SizedBox 组件的主要属性如下所示：

属性名	类型	说明
width	AlignmentGeometry	宽度值，如果具体设置了，则强制 child 宽度为此值；如果没设置，则根据 child 宽度调整自身宽度。
height	double	高度值，如果具体设置了，则强制 child 高度为此值；如果没设置，则根据 child 高度调整自身宽度。

接下来我们写个例子，添加一个 SizedBox 容器，在它里面放一个 Card，那么 Card 就被限定在宽 200、高 300 的范围内。完整的示例代码如下：

```
import 'package:flutter/material.dart';

class LayoutDemo extends StatelessWidget {
  @override
  Widget build(BuildContext context) {
    return new Scaffold(
      appBar: new AppBar(
        title: new Text('SizedBox 设置具体尺寸示例'),
      ),
      body: SizedBox(
        // 固定宽为 200.0，高为 300.0
        width: 200.0,
        height: 300.0,
        child: const Card(
            child: Text('SizedBox',
              style: TextStyle(
                fontSize: 36.0,
              ),
            )),
      ),
    );
  }
}

void main() {
  runApp(
    new MaterialApp(
      title: 'OverflowBox 溢出父容器显示示例',
      home: new LayoutDemo(),
    ),
  );
}
```

上述示例代码的视图展现大致如图 7-28 所示。

图 7-28　SizedBox 组件示例

7.2.2 ConstrainedBox（限定最大最小宽高布局）

ConstrainedBox 的作用是限定子元素 child 的最大宽度、最大高度、最小宽度和最小高度。ConstrainedBox 主要属性如下所示：

属性名	类型	说明
constraints	BoxConstraints	添加到 child 上的额外限制条件，其类型为 BoxConstraints，BoxConstraints 的作用就是限制各种最大最小宽高。
child	Widget	子元素，任意 Widget。

接下来我们写一个小示例，示例中在一个宽高为 300.0 的 Container 上添加一个约束最大最小宽高的 ConstrainedBox，实际显示中，则是一个宽高为 220.0 的区域。示例代码如下：

```
import 'package:flutter/material.dart';
class LayoutDemo extends StatelessWidget {
  @override
  Widget build(BuildContext context) {
    return new Scaffold(
      appBar: new AppBar(
        title: new Text('ConstrainedBox 限定宽高示例'),
      ),
      body: new ConstrainedBox(
        constraints: const BoxConstraints(
          minWidth: 150.0,
          minHeight: 150.0,
          maxWidth: 220.0,
          maxHeight: 220.0,
        ),
        child: new Container(
          width: 300.0,
          height: 300.0,
          color: Colors.green,
        ),
      ),
    );
  }
}
void main() {
  runApp(
    new MaterialApp(
      title: 'ConstrainedBox 限定宽高示例',
      home: new LayoutDemo(),
    ),
  );
}
```

> **注意** 如果 child 不为 null，则将限制条件加在 child 上。如果 child 为 null，则会尽可能地缩小尺寸。

上述示例代码的视图展现大致如图 7-29 所示。

图 7-29　ConstrainedBox 限定宽高示例

7.2.3　LimitedBox（限定最大宽高布局）

LimitedBox 组件是限制类型的组件，可对最大宽高进行限制。和 ConstrainedBox 类似，只不过 LimitedBox 组件没有最小宽高限制。

从布局的角度讲，LimitedBox 是将 child 限制在其设定的最大宽高中的，但这个限定是有条件的。当 LimitedBox 最大宽度不受限制时，child 的宽度就会受到这个最大宽度的限制，高度同理。LimitedBox 组件的主要属性如下所示：

属性名	类型	说明
maxWidth	double	限定的最大宽度，默认值是 double.infinity。
maxHeight	double	限定的最大高度，默认值是 double.infinity。

接下来我们写一个小示例，添加两个容器，第一个容器宽度为 100，高度撑满父容器。第二个容器虽然宽度设置为 250，但是由于它的父容器用 LimitedBox 限定的最大宽度为 150，所以它的实际宽度为 150，高度也同样撑满父容器。完整的示例代码如下：

```
import 'package:flutter/material.dart';
class LayoutDemo extends StatelessWidget {
  @override
```

```
Widget build(BuildContext context) {
  return new Scaffold(
    appBar: new AppBar(
      title: new Text('LimitedBox限定宽高布局示例'),
    ),
    body: Row(
      children: <Widget>[
        Container(
          color: Colors.grey,
          width: 100.0,
        ),
        LimitedBox(
          maxWidth: 150.0,// 设置最大宽度，限定child在此范围内
          child: Container(
            color: Colors.lightGreen,
            width: 250.0,
          ),
        ),
      ],
    )
  );
}
void main() {
  runApp(
    new MaterialApp(
      title: 'LimitedBox限定宽高布局示例',
      home: new LayoutDemo(),
    ),
  );
}
```

上述示例代码的视图展现大致如图 7-30 所示。

7.2.4 AspectRatio（调整宽高比）

AspectRatio 的作用是根据设置调整子元素 child 的宽高比，Flutter 提供此组件，就免去了自定义所带来的麻烦。AspectRatio 适用于需要固定宽高比的情景。

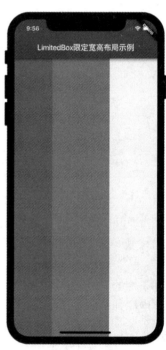

图 7-30 LimitedBox 限定宽高布局示例

AspectRatio 的布局行为分为两种情况：

- AspectRatio 首先会在布局限制条件允许的范围内尽可能地扩展，Widget 的高度是由宽度和比率决定的，类似于 BoxFit 中的 contain，按照固定比率去尽量占满区域。
- 如果在满足所有限制条件后无法找到可行的尺寸，AspectRatio 最终将会优先适应布

局限制条件，而忽略所设置的比率。

AspectRatio 主要属性如下所示：

属性名	类型	说明
aspectRatio	double	宽高比，最终可能不会根据这个值去布局，具体则要看综合因素，外层是否允许按照这种比率进行布局，这只是一个参考值。
child	Widget	子元素，任意 Widget。

接下来我们写一个小示例，示例代码定义了一个高度为 200 的区域，内部 AspectRatio 比率设置为 1.5，最终 AspectRatio 是宽 300 高 200 的一个区域。示例代码如下：

```
import 'package:flutter/material.dart';
class LayoutDemo extends StatelessWidget {
  @override
  Widget build(BuildContext context) {
    return new Scaffold(
      appBar: new AppBar(
        title: new Text('AspectRatio 调整宽高比示例'),
      ),
      body: new Container(
        height: 200.0,
        child: new AspectRatio(
          aspectRatio: 1.5,// 比例可以调整
          child: new Container(
            color: Colors.green,
          ),
        ),
      ),
    );
  }
}
void main() {
  runApp(
    new MaterialApp(
      title: 'AspectRatio 调整宽高比',
      home: new LayoutDemo(),
    ),
  );
}
```

> **注意** aspectRatio 不能为 null，其值必须大于 0 且必须是有限的。

上述示例代码的视图展现大致如图 7-31 所示。

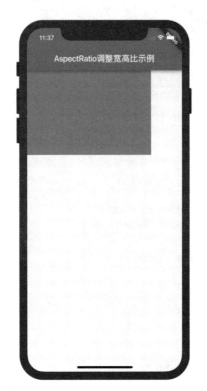

图 7-31　AspectRatio 调整宽高比示例

7.2.5　FractionallySizedBox（百分比布局）

FractionallySizedBox 组件会根据现有空间来调整 child 的尺寸，所以就算为 child 设置了具体的尺寸数值，也不起作用。当需要在一个区域里面取百分比尺寸时，可以使用这个组件，比如需要一个高度 30% 宽度 60% 的区域。

FractionallySizedBox 的布局主要跟它的宽高因子（两个参数）有关，当参数为 null 或者有具体数值的时候，布局表现不一样。当然，还有一个辅助参数 alignment，作为对齐方式进行布局。具体有以下两种情况：

❑ 设置了具体的宽高因子，具体的宽高则根据现有空间宽高 × 因子，当宽高因子大于 1 的时候，有可能会超出父组件的范围。
❑ 没有设置宽高因子，则填满可用区域。

FractionallySizedBox 组件的主要属性如下所示：

属性名	类型	说明
alignment	AlignmentGeometry	对齐方式，不能为 null。
widthFactor	double	宽度因子，宽度乘以这个值，就是最后的宽度。
heightFactor	double	高度因子，用作计算最后实际高度的。

> **注意**：如果宽高因子不为 null，那么实际的最大宽高度则为 child 的宽高乘以这个因子；如果因子为 null，那么 child 的宽高则会尽量充满整个区域。

接下来我们写一个小示例，示例中有两个容器，底部的容器宽高各为 200。如果上面的容器宽度因子为 0.5，则它的实际宽度为 100；如果高度因子为 1.5，则它的实际高度为 300。对齐方式为左上角对齐。所以上层的元素横向在容器内部，纵向已经超出容器了。完整的示例代码如下：

```
import 'package:flutter/material.dart';
class LayoutDemo extends StatelessWidget {
  @override
  Widget build(BuildContext context) {
    return new Scaffold(
      appBar: new AppBar(
        title: new Text('FractionallySizedBox百分比布局示例'),
      ),
      body: new Container(
        color: Colors.blueGrey,
        height: 200.0,
        width: 200.0,
        child: new FractionallySizedBox(
          alignment: Alignment.topLeft,// 元素左上角
                                       对齐

          widthFactor: 0.5,// 宽度因子
          heightFactor: 1.5,// 高度因子
          child: new Container(
            color: Colors.green,
          ),
        ),
      ),
    );
  }
}
void main() {
  runApp(
    new MaterialApp(
      title: 'FractionallySizedBox百分比布局示例',
      home: new LayoutDemo(),
    ),
  );
}
```

上述示例代码的视图展现大致如图 7-32 所示。

图 7-32　FractionallySizedBox 组件示例

7.3　列表及表格布局

Flutter 中列表和表格的布局方式有多种，本节将介绍这类组件。

7.3.1 ListView

ListView 布局是一种常用的布局方式，ListView 结合 ListTitle 可以布局出一些复杂的列表界面。

具体示例代码如下：

```
import 'package:flutter/material.dart';

void main() {
  runApp(new MaterialApp(
    title: 'ListView 布局示例 ',
    home: new MyApp(),
  ));
}

class MyApp extends StatelessWidget {
  @override
  Widget build(BuildContext context) {

    List<Widget> list = <Widget>[
      new ListTile(
        title: new Text(' 广州市黄埔大道建中路店 ',style: new TextStyle(fontWeight:
          FontWeight.w400,fontSize: 18.0),),
        subtitle: new Text(' 广州市福黄埔大道建中路 3 号 '),
        leading: new Icon(
          Icons.fastfood,
          color: Colors.orange,
        ),
      ),
      new ListTile(
        title: new Text(' 广州市白云区机场路白云机场店 ',style: new
          TextStyle(fontWeight: FontWeight.w400,fontSize: 18.0),),
        subtitle: new Text(' 广州市白云区机场路 T3 航站楼 '),
        leading: new Icon(
          Icons.airplay,
          color: Colors.blue,
        ),
      ),
      new ListTile(
        title: new Text(' 广州市中山大道中山大学附属医院 ',style: new
          TextStyle(fontWeight: FontWeight.w400,fontSize: 18.0),),
        subtitle: new Text(' 广州市中山大道 45 号 '),
        leading: new Icon(
          Icons.local_hospital,
          color: Colors.green,
        ),
      ),
      new ListTile(
        title: new Text(' 广州市天河区太平洋数码城 ',style: new TextStyle(fontWeight:
```

```
        FontWeight.w400,fontSize: 18.0),),
      subtitle: new Text('广州市天河区岗顶太平洋数码城'),
      leading: new Icon(
        Icons.computer,
        color: Colors.deepPurple,
      ),
    ),
  ];


  return new Scaffold(
    appBar: new AppBar(
      title: new Text('ListView布局示例'),
    ),
    body: new Center(
      child: new ListView(
        children: list,
      ),
    ),
  );
}
```

上述示例代码的视图展现大致如图7-33所示。

ListView还可以实现长文本的滚动效果,一般可用于页面内容较多的场景。示例代码如下:

```
import 'package:flutter/material.dart';

void main() {
  runApp(new MaterialApp(
    title: '滚动布局示例',
    home: new MyApp(),
  ));
}

class MyApp extends StatelessWidget {
  @override
  Widget build(BuildContext context) {
```

图 7-33　ListView 布局示例

```
    return new Scaffold(
      appBar: new AppBar(
        title: new Text('滚动布局示例'),
      ),
      body:new ListView(
        children: <Widget>[
          new Center(
```

```
            child: new Text(
              '\n广州天河区公园',
              style: new TextStyle(
                fontSize:30.0,
              ),
            ),
          ),
          new Center(
            child: new Text(
              '天河公园',
              style: new TextStyle(
                fontSize: 16.0,
              ),
            ),
          ),
          new Center(
            child: new Text(
              '''天河公园，是区属综合性公园，位于广州天河区员村，西靠天府路，南连黄埔大道，北接中山大道，来往交通十分便利。公园总面积为70.7公顷，水体面积占10公顷。天河公园以自然生态景观为主要特色，公园规划为五个功能区：百花园景区、文体娱乐区、老人活动区、森林休憩区、后勤管理区。
              ''',
              style: new TextStyle(
                fontSize: 14.0,
              ),
            ),
          ),
        ],
      ),
    );
  }
}
```

上述示例代码的视图展现大致如图7-34所示。　　图7-34　滚动布局应用示例

 在长文本的场景，需要使用 ''' 三个引号来表示。

7.3.2　GridView

GridView布局即为网格布局，通常用来布局多行多列的情况。接下来编写一个GridView布局的九宫格小例子。具体步骤如下：

1）在helloworld工程下需要添加9张图片，具体添加过程请参考7.1.4节Align对齐布局添加步骤。

2）编写GridView九宫格示例代码，完整的示例代码如下：

```
import 'package:flutter/material.dart';
```

```
void main() {
  runApp(new MaterialApp(
    title: 'GridView 九宫格示例',
    home: new MyApp(),
  ));
}

class MyApp extends StatelessWidget {
  @override
  Widget build(BuildContext context) {
    // 使用 generate 构造图片列表
    List<Container> _buildGridTitleList(int count) {
      return new List<Container>.generate(
          count,
          (int index) => new Container(
            child: new Image.asset('images/${index + 1}.jpeg'),
          ));
    }

    // 渲染 GridView
    Widget buildGrid(){
      return new GridView.extent(
        maxCrossAxisExtent: 150.0,// 次轴的宽度
        padding: const EdgeInsets.all(4.0),// 上下左右内边距
        mainAxisSpacing: 4.0,// 主轴间隙
        crossAxisSpacing: 4.0,// 次轴间隙
        children: _buildGridTitleList(9),// 添加 9 个元素
      );
    }

    return new Scaffold(
      appBar: new AppBar(
        title: new Text('GridView 九宫格示例'),
      ),
      body: new Center(
        child: buildGrid(),
      ),
    );
  }
}
```

上述示例代码的视图展现大致如图 7-35 所示。

7.3.3 Table

几乎每一个前端技术的布局中都会有一种 table 布局，这种组件太常见了，以至于其表现形式，完全可以借鉴其他前端技术。表格布局中，每一行的高度由其内容决定，

图 7-35　GridView 九宫格示例

每一列的宽度由 columnWidths 属性单独控制。

Table 组件属性见表 7-2。

表 7-2 Table 组件属性及描述

属性名	类型	说明
columnWidths	Map<int, TableColumnWidth>	设置每一列的宽度
defaultColumnWidth	TableColumnWidth	默认的每一列宽度值，默认情况下均分
textDirection	TextDirection	文字方向，一般无需考虑
border	TableBorder	表格边框
defaultVerticalAlignment	TableCellVerticalAlignment	默认垂直方向对齐方式： ● top：放置在顶部 ● middle：垂直居中 ● bottom：放置在底部 ● baseline：文本 baseline 对齐 ● fill：充满整个 cell
textBaseline	TextBaseline	defaultVerticalAlignment 为 baseline 的时候，会用到这个属性

接下来编写一个用于员工基本信息统计的表格。Table 表格布局完整的示例代码如下：

```
import 'package:flutter/material.dart';

void main() {
  runApp(new MaterialApp(
    title: 'Table表格布局示例',
    home: new MyApp(),
  ));
}

class MyApp extends StatelessWidget {
  @override
  Widget build(BuildContext context) {

    return new Scaffold(
      appBar: new AppBar(
        title: new Text('Table表格布局示例'),
      ),
      body: new Center(
        child: Table(
          // 设置表格有多少列，并且指定列宽
          columnWidths: const <int, TableColumnWidth>{
            0: FixedColumnWidth(100.0),
            1: FixedColumnWidth(40.0),
            2: FixedColumnWidth(80.0),
            3: FixedColumnWidth(80.0),
          },
          // 设置表格边框样式
```

```
        border: TableBorder.all(color: Colors.black38, width: 2.0, style:
          BorderStyle.solid),
        children: const <TableRow>[
          // 添加第一行数据
          TableRow(
            children: <Widget>[
              Text(' 姓名 '),
              Text(' 性别 '),
              Text(' 年龄 '),
              Text(' 身高 '),
            ],
          ),
          // 添加第二行数据
          TableRow(
            children: <Widget>[
              Text(' 张三 '),
              Text(' 男 '),
              Text('26'),
              Text('172'),
            ],
          ),
          // 添加第三行数据
          TableRow(
            children: <Widget>[
              Text(' 李四 '),
              Text(' 男 '),
              Text('28'),
              Text('178'),
            ],
          ),
        ],
      ),
    );
  }
}
```

上述示例代码的视图展现大致如图 7-36 所示。

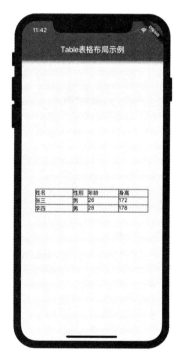

图 7-36 Table 表格布局示例

7.4 其他布局处理

本节介绍的组件包括矩阵转换、基准线布局、组件显示、自动换行等。

7.4.1 Transform（矩阵转换）

Transform 的主要作用就是做矩阵变换。Container 中矩阵变换就使用了 Transform。Transform 可以对 child 做平移、旋转及缩放等操作。

Transform 组件的主要属性如下所示：

属性名	类型	说明
transform	Matrix4	一个 4×4 的矩阵。不难发现，其他平台的变换矩阵也都是四维的。一些复合操作，仅靠三维是不够的，必须采用额外的一维来补充。
origin	Offset	旋转点，相对于左上角顶点的偏移。默认旋转点是左上角顶点。
alignment	AlignmentGeometry	对齐方式。
transformHitTests	bool	点击区域是否也做相应的改变。

接下来编写一个例子，例子中使容器旋转了一定的角度。并且用的是 Matrix4.rotationZ(0.3) 方法进行旋转。完整的示例代码如下：

```
import 'package:flutter/material.dart';

class LayoutDemo extends StatelessWidget {
  @override
  Widget build(BuildContext context) {
    return new Scaffold(
      appBar: new AppBar(
        title: new Text('Transform矩阵转换示例'),
      ),
      body: new Center(
        child: Container(
          color: Colors.grey,
          child: Transform(
            alignment: Alignment.topRight,
            transform: Matrix4.rotationZ(0.3),
            child: Container(
              padding: const EdgeInsets.all(8.0),
              color: const Color(0xFFE8581C),
              child: const Text('Transform矩阵转换'),
            ),
          ),
        ),
      ),
    );
  }
}

void main() {
  runApp(
    new MaterialApp(
      title: 'Transform矩阵转换示例',
      home: new LayoutDemo(),
    ),
  );
}
```

上述示例代码的视图展现大致如图 7-37 所示。

图 7-37 Transform 矩阵转换示例

7.4.2 Baseline（基准线布局）

Baseline 基准线是指将所有元素底部放在同一条水平线上。做过移动端开发的读者大都了解过，一般文字排版的时候可能会用到它。它的作用很简单，根据 child 的 baseline 来调整 child 的位置。例如两个字号不一样的文字，希望底部在一条水平线上，就可以使用这个组件，是一个非常基础的组件。

图 7-38 中两个字线一上一下，如果想对齐，就需要使用到 Baseline 组件来控制。

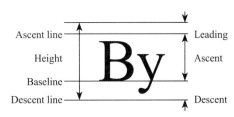

图 7-38　Baseline 组件示意图

Baseline 组件的主要属性如下所示：

属性名	类型	说明
baseline	double	baseline 数值，必须要有，从顶部算。
baselineType	TextBaseline	baseline 类型，也是必须要有的，目前有两种类型。alphabetic 对齐字符底部的水平线，ideographic 对齐表意字符的水平线。

接下来编写一个小示例，示例中三个元素在同一个水平线上。baseline 也设定相同的值。示例代码如下：

```
import 'package:flutter/material.dart';
class LayoutDemo extends StatelessWidget {
  @override
  Widget build(BuildContext context) {
    return new Scaffold(
      appBar: new AppBar(
        title: new Text('Baseline 基准线布局示例'),
      ),
      body: new Row(
        // 水平等间距排列子组件
        mainAxisAlignment: MainAxisAlignment.spaceBetween,
        children: <Widget>[
          // 设置基准线
          new Baseline(
            baseline: 80.0,
            // 对齐字符底部水平线
            baselineType: TextBaseline.alphabetic,
            child: new Text(
              'AaBbCc',
              style: new TextStyle(
                fontSize: 18.0,
                textBaseline: TextBaseline.alphabetic,
              ),
            ),
          ),
```

```
        new Baseline(
          baseline: 80.0,
          baselineType: TextBaseline.alphabetic,
          child: new Container(
            width: 40.0,
            height: 40.0,
            color: Colors.green,
          ),
        ),
        new Baseline(
          baseline: 80.0,
          baselineType: TextBaseline.alphabetic,
          child: new Text(
            'DdEeFf',
            style: new TextStyle(
              fontSize: 26.0,
              textBaseline: TextBaseline.
                alphabetic,
            ),
          ),
        ),
      ],
    ),
  );
  }
}
void main() {
  runApp(
    new MaterialApp(
      title: 'Baseline基准线布局示例',
      home: new LayoutDemo(),
    ),
  );
}
```

上述示例代码的视图展现大致如图 7-39 所示。

7.4.3　Offstage（控制是否显示组件）

Offstage 的作用很简单，通过一个参数来控制 child 是否显示，也算是比较常用的组件。Offstage 组件的主要属性如下所示：

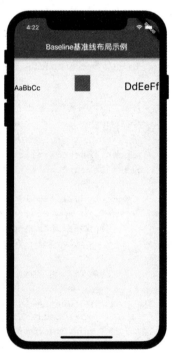

图 7-39　Baseline 基准线布局示例

属性名	类型	默认值	说明
offstage	bool	true	默认为 true 表示不显示，当为 false 的时候，会显示该组件。

这里编写一个控制文本显示隐藏的小示例，需要添加一个控制状态的变量 offstage。示例里点击右下角按钮可以显示或隐藏文本内容。完整的示例代码如下：

```
import 'package:flutter/material.dart';

void main() {
  runApp(new MyApp());
}

class MyApp extends StatelessWidget {

  @override
  Widget build(BuildContext context) {
    final appTitle = "Offstage 控制是否显示组件示例";
    return new MaterialApp(
      title: appTitle,
      home: new MyHomePage(title:appTitle),
    );
  }
}

class MyHomePage extends StatefulWidget {
  final String title;

  MyHomePage({Key key,this.title}):super(key:key);

  @override
  _MyHomePageState createState() => new _MyHomePageState();

}

class _MyHomePageState extends State<MyHomePage> {

  // 状态控制是否显示文本组件
  bool offstage = true;

  @override
  Widget build(BuildContext context) {
    return new Scaffold(
      appBar: new AppBar(
        title: new Text(widget.title),
      ),
      body: Center(
        child: new Offstage(
          offstage: offstage,// 控制是否显示
          child: new Text(
            '我出来啦!',
            style: TextStyle(
              fontSize: 36.0,
            ),
          ),
        ),
      ),
```

```
      floatingActionButton: new FloatingActionButton(
        onPressed: (){

          //设置是否显示文本组件
          setState(() {
            offstage = !offstage;
          });

        },
        tooltip: "显示隐藏",
        child: new Icon(Icons.flip),
      ),

    );
  }

}
```

上述示例代码的视图展现大致如图 7-40 所示。

图 7-40　Offstage 组件布局示例

7.4.4　Wrap（按宽高自动换行布局）

Wrap 使用了 Flex 中的一些概念，某种意义上说跟 Row、Column 更加相似。单行的 Wrap 跟 Row 表现几乎一致，单列的 Wrap 则跟 Column 表现几乎一致。但 Row 与 Column 都是单行单列的，Wrap 却突破了这个限制，主轴上空间不足时，则向次轴上去扩展显示。对于一些需要按宽度或高度让 child 自动换行布局的场景，可以使用 Wrap。

Wrap 组件属性见表 7-3。

表 7-3　Wrap 组件属性及描述

属性名	类型	默认值	说明
direction	Axis	Axis.horizontal	主轴（mainAxis）的方向，默认为水平
alignment	WrapAlignment		主轴方向上的对齐方式，默认为 start
spacing	double	0.0	主轴方向上的间距
runAlignment	WrapAlignment	WrapAlignment.start	run 的对齐方式。run 可以理解为新的行或者列，如果是水平方向布局的话，run 可以理解为新的一行
runSpacing	double	0.0	run 的间距
crossAxisAlignment	WrapCrossAlignment	WrapCrossAlignment.start	主轴（crossAxis）方向上的对齐方式
textDirection	TextDirection		文本方向
verticalDirection	VerticalDirection		定义了 children 摆放顺序，默认是 down，见 Flex 相关属性介绍

接下来编写一个例子，在容器里放一些头像。使用 Wrap 组件包装以后，头像可以按从左到右和从上到下的顺序排列头像元素。Wrap 布局的完整示例代码如下：

```
import 'package:flutter/material.dart';

void main() {
  runApp(new MaterialApp(
    title: 'Wrap按宽高自动换行布局示例',
    home: new MyApp(),
  ));
}

class MyApp extends StatelessWidget {
  @override
  Widget build(BuildContext context) {

    return new Scaffold(
      appBar: new AppBar(
        title: new Text('Wrap按宽高自动换行布局示例'),
      ),
      body: Wrap(
        spacing: 8.0, // Chip 之间的间隙大小
        runSpacing: 4.0, // 行之间的间隙大小
        children: <Widget>[
          Chip(
            // 添加圆形头像
            avatar: CircleAvatar(
              backgroundColor: Colors.lightGreen.shade800, child: new Text('西门', style: TextStyle(fontSize: 10.0),)),
```

```
              label: Text('西门吹雪'),
            ),
            Chip(
              avatar: CircleAvatar(
                  backgroundColor: Colors.lightBlue.shade700, child: new Text('司
                      空', style: TextStyle(fontSize: 10.0),)),
              label: Text('司空摘星'),
            ),
            Chip(
              avatar: CircleAvatar(
                  backgroundColor: Colors.orange.shade800, child: new Text('婉清',
                      style: TextStyle(fontSize: 10.0),)),
              label: Text('木婉清'),
            ),
            Chip(
              avatar: CircleAvatar(
                  backgroundColor: Colors.blue.shade900, child: new Text('一郎',
                      style: TextStyle(fontSize: 10.0),)),
              label: Text('萧十一郎'),
            ),
          ],
        ),
      );
    }
}
```

上述示例代码的视图展现大致如图 7-41 所示。

图 7-41　Wrap 自动换行布局示例

7.5 布局综合示例

上面讲了这么多布局的组件，还没有做一个完整的页面。这里我们通过一个风景区的介绍来讲解布局的综合运用。先上效果图，如图 7-42 所示。

图 7-42 布局综合示例效果图

7.5.1 布局分析

整体布局使用一个垂直布局组件 ListView 进行滚动布局。一共有四大块：武当山图片、风景区地址、按钮组和景区介绍文本块。整体拆分如图 7-43 所示，共计四个方框。

> **注意** 垂直方向使用 ListView 而不使用 Column 的原因是，风景区介绍文本可能会很长，甚至超出屏幕，如果使用 Column 组件，则有部分文本可能会看不到。使用 ListView 组件，用户可以滚动到下面查看文本。

接下来细拆分这四大块。最上面图片及最下面文本块由于是单个组件即可完成，不需要细拆。所以重点介绍风景区地址及按钮组。

图 7-43　总体拆分图

风景区地址从横向上看需要使用一个水平排列的组件 Row，水平方向总共有三个 Child，分别为左侧文本区域、右侧图标及数字区域。如图 7-44 所示。其中左侧文本区域要继续细拆，需要用一个垂直布局的组件 Column，上下各放一个文本组件即可。右侧图标及数字是两个组件，所以横向上来看总共是三个组件。

这里有个问题，左侧及右侧之间的空隙怎么解决？如图 7-45 方框区域所示。这里需要用 Expanded 组件来包裹风景区地址组件以达到填充空隙的目的。

图 7-44　风景区地址水平方向拆分　　　　图 7-45　风景区地址中间空隙

接下来分析按钮组的布局，在横向上用 Row 组件排列三个按钮。在纵向上用 Column 做三个相同的按钮，上面为按钮图标，下面为按钮文本。这样布局拆分的好处是，最大化地复用组件。具体拆分如图 7-46 所示。

图 7-46　按钮组拆分示意图

7.5.2　准备素材

这里还是使用 helloworld 工程，在工程的 images 目录下添加一张风景图片叫 wudang.jpeg。如图 7-47 所示。

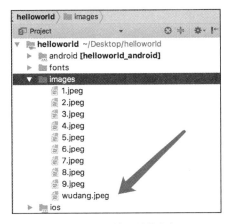

图 7-47　添加风景图片

打开工程根目录下的 pubspec.yaml 文件，在 assets 配置选项下添加 -images/wudang.jpeg。添加好后，点击上面的 Packages get 进行资源的更新。如图 7-48 所示。

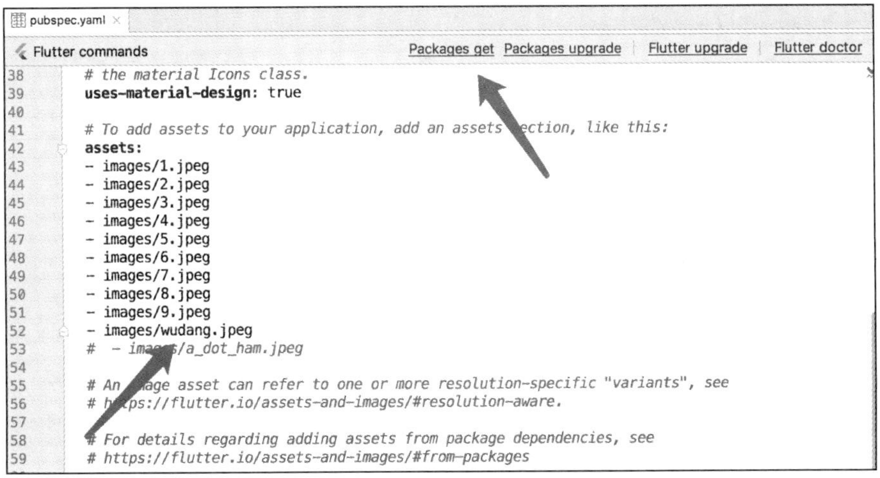

图 7-48　修改 pubspec.yaml 属性值

注意　图片资源路径及文件名一定不要写错，否则运行时会报错。点击 **Packages get** 选项更新资源即可，不要点击其他选项，有可能由升级带来其他的问题。

7.5.3 编写代码

1. 图片处理

图片的宽度尽量大一些，填充方式使用 BoxFit.cover 模式，这样可以填充整个父容器，如果不铺满则会显得很不美观。具体代码如下所示：

```
Image.asset(
  'images/wudang.jpeg',
  width: 600.0,
  height: 240.0,
  fit: BoxFit.cover, // 图片填充整个父容器
),
```

2. 风景地址处理

风景区地址部分相对复杂，首先横向添加了一个 Row 组件，然后用了一个 Expanded 组件包裹左侧文本区域，以便填充左右组件的空隙部分。在左侧部分又添加了一个垂直布局 Column，用来放置两行文本，文字之间加了一个 8.0 的间距。最后右侧添加图标及数字组件。此部分的代码结构如下所示：

```
child: Row(
  children: <Widget>[
    Expanded(
      child: Column(
        crossAxisAlignment// 次轴对齐方式
        children: <Widget>[
          Container(
            padding// 与下面文本间隔一定距离
            // 添加标题文本
          ),
          // 添加地址文本
        ],
      ),
    ),
    // 右侧图标及数字
  ],
),
);
```

3. 按钮组处理

首先编写一个方法，用来构建单个按钮，需要传入图标及文本这两个参数，采用垂直布局方式。代码如下所示：

```
Column buildButtonColumn(IconData icon, String label) {
```

```
    return Column(
       // 按钮界面渲染 图标+文本
    );
}
```

组装三个按钮,首先添加一个水平布局 Row,水平方向对齐方式采用均匀排列方式。然后调用 buildButtonColumn 方法传入三组数据,构造三个按钮。代码结构如下所示:

```
Widget buttonsContainer = Container(
    child: Row(
    mainAxisAlignment// 水平方向对齐方式
    children: <Widget>[
      buildButtonColumn// 构建三个按钮
    ],
    ),
);
```

4. 风景区介绍文本部分

风景区介绍文本部分实现起来相对容易,只需要添加一个 Text 组件即可,需要注意大的文本块要使用三个单引号引用起来。具体代码如下所示:

```
child: Text(
    '''
    大文本块
    ''',
    softWrap: true,
),
```

5. 自定义主题

由于是风景区的介绍,通常希望颜色以绿色风格为主,所以这里需要自定义主题。具体代码如下所示:

```
theme: new ThemeData(
    // 设置主题各个颜色值
),
```

6. 组装所有内容

将上面的所有内容组装起来,采用 ListView 组件以避免文本块过长导致无法查看下面文本的问题。代码结构如下:

```
return new MaterialApp(
    title// 标题
    theme// 自定义主题,
    home: Scaffold(
      appBar: AppBar(
      ),
      body: ListView(
        children: <Widget>[
```

```
          // 景区图片
          // 风景区地址
          // 按钮组
          // 风景区介绍文本
        ],
      ),
    ),
  );
```

布局示例的完整代码如下所示：

```
import 'package:flutter/material.dart';

void main() => runApp(new MyApp());

class MyApp extends StatelessWidget {
  @override
  Widget build(BuildContext context) {
    // 风景区地址部分
    Widget addressContainer = Container(
      padding: const EdgeInsets.all(32.0),// 此部分四周间隔一定距离
      child: Row(
        children: <Widget>[
          Expanded(
            child: Column(
              crossAxisAlignment: CrossAxisAlignment.start, // 顶部对齐
              children: <Widget>[
                Container(
                  padding: const EdgeInsets.only(bottom: 8.0),// 与下面文本间隔一定距离
                  child: Text(
                    '风景区地址',
                    style: TextStyle(
                      fontWeight: FontWeight.bold,
                    ),
                  ),
                ),
                Text(
                  '湖北省十堰市丹江口市',
                  style: TextStyle(
                    color: Colors.grey[500],
                  ),
                ),
              ],
            ),
          ),
          Icon(
            Icons.star,
            color: Colors.red[500],
          ),
          Text('66'),
        ],
```

```
    ),
  );

  // 构建按钮组中单个按钮 参数为图标及文本
  Column buildButtonColumn(IconData icon, String label) {
    return Column(
      mainAxisSize: MainAxisSize.min,// 垂直方向大小最小化
      mainAxisAlignment: MainAxisAlignment.center,// 垂直方向居中对齐
      children: <Widget>[
        Icon(icon, color: Colors.lightGreen[600]),// 上面图标部分
        Container(
          margin: const EdgeInsets.only(top: 8.0),
          child: Text(// 下面文本部分
            label,
            style: TextStyle(
              fontSize: 12.0,
              fontWeight: FontWeight.w400,
              color: Colors.lightGreen[600],
            ),
          ),
        ),
      ],
    );
  }

  // 按钮组部分
  Widget buttonsContainer = Container(
    // 容器横向布局
    child: Row(
      mainAxisAlignment: MainAxisAlignment.spaceEvenly,// 水平方向均匀排列每个元素
      children: <Widget>[
        buildButtonColumn(Icons.call, '电话'),
        buildButtonColumn(Icons.near_me, '导航'),
        buildButtonColumn(Icons.share, '分享'),
      ],
    ),
  );

  // 风景区介绍文本部分
  Widget textContainer = Container(
    padding: const EdgeInsets.all(32.0),
    // 文本块一定是用 ''' 来引用起来
    child: Text(
      '''
      武当山，中国道教圣地，又名太和山、谢罗山、参上山、仙室山，古有"太岳""玄岳""大岳"之
      称。位于湖北西北部十堰市丹江口市境内。东接闻名古城襄阳市，西靠车城十堰市 ，南望原始森林神
      农架，北临高峡平湖 丹江口水库。
      明代，武当山被皇帝封为"大岳"、"治世玄岳"，被尊为"皇室家庙"。武当山以"四大名山皆拱揖，
      五方仙岳共朝宗"的"五岳之冠"地位闻名于世。
      1994年12月，武当山古建筑群入选《世界遗产名录》，2006年被整体列为"全国重点文物保
```

护单位"。2007 年，武当山和长城、丽江、周庄等景区一起入选"欧洲人最喜爱的中国十大景区"。
2010 至 2013 年，武当山分别被评为国家 5A 级旅游区、国家森林公园、中国十大避暑名山、海峡两
岸交流基地，入选最美"国家地质公园"。

截至 2013 年，武当山有古建筑 53 处，建筑面积 2.7 万平方米，建筑遗址 9 处，占地面积 20 多万
平方米，全山保存各类文物 5035 件。

武当山是道教名山和武当武术的发源地，被称为"亘古无双胜境，天下第一仙山"。武当武术，是中
华武术的重要流派。元末明初，道士张三丰集其大成，开创武当派。

```
      ''',
      softWrap: true,
    ),
  );

  return new MaterialApp(
    title: '布局综合示例',
    // 自定义主题，主体颜色为绿色风格
    theme: new ThemeData(
      brightness: Brightness.light, // 应用程序整体主题的亮度
      primaryColor: Colors.lightGreen[600], //App 主要部分的背景色
      accentColor: Colors.orange[600], // 前景色（文本、按钮等）
    ),
    home: Scaffold(
      appBar: AppBar(
        title: Text(
          '武当山风景区',
          style: TextStyle(color: Colors.white),
        ),
      ),
      body: ListView(
        children: <Widget>[
          Image.asset(
            'images/wudang.jpeg',
            width: 600.0,
            height: 240.0,
            fit: BoxFit.cover, // 图片填充整个父容器
          ),
          addressContainer,
          buttonsContainer,
          textContainer,
        ],
      ),
    ),
  );
}
}
```

第 8 章 Chapter 8

手 势

移动应用中一个必不可少的环节就是与用户的交互，在 Android 中提供了手势检测，并为手势检测提供了相应的监听。Flutter 中提供的手势检测为 GestureDetector。

Flutter 中的手势系统分为两层，第一层是触摸原事件（指针），有相应的四种事件类型：
- PointerDownEvent：用户与屏幕接触产生了联系。
- PointerMoveEvent：手指已从屏幕上的一个位置移动到另一个位置。
- PointerUpEvent：用户已停止接触屏幕。
- PointerCancelEvent：此指针的输入不再指向此应用程序。

第二层就是我们可以检测到的手势，主要分为三大类，包括轻击、拖动和缩放。

本章通过介绍手势的处理方法，来展示 Flutter 手势的基本用法。内容如下：
- 用 GestureDetector 进行手势检测
- 用 Dismissible 实现滑动删除

8.1 用 GestureDetector 进行手势检测

GestureDetector 可以进行手势检测，比如点击一次、双击、长按、垂直拖动及水平拖动等。这些手势的事件名及描述如表 8-1 所示。

表 8-1 GestureDetector 事件描述

事件名	描述
onTapDown	点击屏幕立即触发此方法
onTapUp	手指离开屏幕

（续）

事件名	描述
onTap	点击屏幕
onTapCancel	此次点击事件结束，onTapDown 不会在产生点击事件
onDoubleTap	用户快速连续两次在同一位置点击屏幕
onLongPress	长时间保持与相同位置的屏幕接触
onVerticalDragStart	与屏幕接触，可能会开始垂直移动
onVerticalDragUpdate	与屏幕接触并垂直移动的指针在垂直方向上移动
onVerticalDragEnd	之前与屏幕接触并垂直移动的指针不再与屏幕接触，并且在停止接触屏幕时以特定速度移动垂直拖动
onHorizontalDragStart	与屏幕接触，可能开始水平移动
onHorizontalDragUpdate	与屏幕接触并水平移动的指针在水平方向上移动
onHorizontalDragEnd	先前与屏幕接触并且水平移动的指针不再与屏幕接触，并且当它停止接触屏幕时以特定速度移动水平拖动

假设我们想要制作一个自定义按钮，当点击时显示文字"你已按下"。我们如何解决这个问题？请看下面示例代码：

```
import 'package:flutter/material.dart';

void main() {
  runApp(new MaterialApp(
    title: '按下处理 Demo',
    home: new MyApp(),
  ));
}

class MyButton extends StatelessWidget{
  @override
  Widget build(BuildContext context) {

    // 一定要把被触摸的组件放在 GestureDetector 里面
    return new GestureDetector(
      onTap: (){
        // 底部消息揭示
        final snackBar = new SnackBar(content: new Text(" 你已按下 "),);
        Scaffold.of(context).showSnackBar(snackBar);
      },
      // 添加容器接收触摸动作
      child: new Container(
        padding: new EdgeInsets.all(12.0),
        decoration: new BoxDecoration(
          color: Theme.of(context).buttonColor,
          borderRadius: new BorderRadius.circular(10.0),
        ),
        child: new Text(' 测试按钮 '),
      ),
```

```
      );
    }
  }

class MyApp extends StatelessWidget {
    @override
    Widget build(BuildContext context) {

      return new Scaffold(
        appBar: new AppBar(
          title: new Text('按下处理 Demo'),
        ),
        body:new Center(child: new MyButton(),)
      );
    }
  }
```

上述示例代码展示总共有两张图，图 8-1 为用户没有点击测试按钮，当用户点击了测试按钮后，会弹出一个提示"你已按下"，如图 8-2 所示。

图 8-1　未点击测试按钮

图 8-2　点击测试按钮后

8.2　用 Dismissible 实现滑动删除

滑动删除模式在很多移动应用中很常见。例如，我们在整理手机通讯录时，希望能快速删除一些联系人，一般用手指轻轻一滑即可以实现删除功能。Flutter 通过提供 Dismissible 组件使这项任务变得简单。Dismissible 组件属性见表 8-2。

表 8-2　Dismissible 组件属性及描述

属性名	类型	说明
onDismissed	DismissDirectionCallback	当包裹的组件消失后回调的函数
movementDuration	Duration	定义组件消息的时长
onResize	VoidCallback	组件大小改变时回调的函数
resizeDuration	Duration	组件大小改变时长
child	Widget	组件包裹的子元素，即被隐藏的对象

接下来编写一个示例，可以删除列表中的某一项数据。其中关键的部分是列表项需要使用 Dismissible 包裹。完整的示例代码如下：

```
import 'package:flutter/material.dart';

void main() {
  runApp(new MaterialApp(
    title: '滑动删除示例',
    home: new MyApp(),
  ));
}

class MyApp extends StatelessWidget {
  // 构建30条列表数据
  List<String> items = new List<String>.generate(30, (i) => "列表项 ${i + 1}");

  @override
  Widget build(BuildContext context) {
    return new Scaffold(
      appBar: new AppBar(
        title: new Text('滑动删除示例'),
      ),
      // 构建列表
      body: new ListView.builder(
        itemCount: items.length,// 指定列表长度
        itemBuilder: (context, index) {// 构建列表

          // 提取出被删除的项
          final item = items[index];

          // 返回一个可以被删除的列表项
          return new Dismissible(
            key: new Key(item),
            // 被删除回调
            onDismissed: (direction) {
              // 移除指定索引项
```

```
            items.removeAt(index);
            // 底部弹出消息提示当前项被删除了
            Scaffold.of(context).showSnackBar(
                new SnackBar(content: new Text("$item 被删除了")));
          },
          child: new ListTile(title: new Text('$item'),)
        );
      },
    ),
  );
}
```

上述示例代码展示了一个列表，如图 8-3 所示，其中"列表项 5"还在。

当用户滑动删除了此项，屏幕底部会弹出一个提示"列表项 5 被删除了"，如图 8-4 所示。

图 8-3 未删除列表项前

图 8-4 滑动删除列表项后

Chapter 9 第 9 章

资源和图片

Flutter 应用程序可以包含代码和 assets（资源）。asset 打包到程序安装包中，可在运行时访问。常见类型的 asset 包括静态数据（例如 JSON 文件）、配置文件、图标和图片（JPEG、WebP、GIF、动画 WebP / GIF、PNG、BMP 和 WBMP）。

手机的资源是多种多样的，但都少不了加载、处理及展示这几个环节。本章通过以下两个方面来给大家讲解资源是如何使用的：

- 添加资源和图片
- 自定义字体

9.1 添加资源和图片

9.1.1 指定 assets

Flutter 使用 pubspec.yaml 文件（位于项目根目录，如：helloworld/pubspec.yaml。文档地址：https://www.dartlang.org/tools/pub/pubspec），来识别应用程序所需的资源 asset。

配置内容如下所示：

```
flutter:

  # The following line ensures that the Material Icons font is
  # included with your application, so that you can use the icons in
  # the material Icons class.
  uses-material-design: true

  # To add assets to your application, add an assets section, like this:
```

```
assets:
 - images/1.jpeg
 - images/2.jpeg
 - images/3.jpeg
 - images/4.jpeg
 - images/5.jpeg
 - images/6.jpeg
 - images/7.jpeg
 - images/8.jpeg
 - images/9.jpeg
 - images/wudang.jpeg
 - images/code.jpeg
```

配置完后，需要点击 Packages get 以更新资源，如图 9-1 所示。

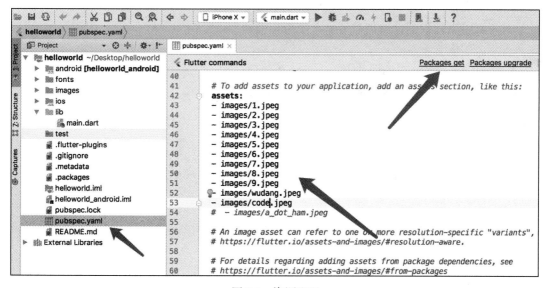

图 9-1 资源配置

> 注意 这里的配置一定要和程序中指定的路径相对应，否则会报找不到资源的错误。应用里所有的静态资源都配置在这里，注意命名规范。

9.1.2 加载 assets

你的应用可以通过 AssetBundle 对象访问其 asset。具体方法如下。

1. 加载文本配置文件

每个 Flutter 应用程序都有一个 rootBundle 对象，可以轻松访问主资源包。可以直接使用 package:flutter/services.dart 中全局静态的 rootBundle 对象来加载资源。

一个应用免不了有一些配置文件，比如放置数据库连接信息、服务器地址、应用名称及版本等。当应用程序启动时往往需要把这些配置项加载出来。

下面的示例就是如何使用 rootBundle 来加载 config.json 配置文件：

```
import 'dart:async' show Future;
import 'package:flutter/services.dart' show rootBundle;

Future<String> loadAsset() async {
  return await rootBundle.loadString('assets/config.json');
}
```

 加载配置文件需要使用异步处理，因为文件的读写操作需要时间，并且不可预知。

2. 加载图片

Flutter 可以为当前设备加载适合其分辨率的图像。不同的手机 DPI 是不同的，DPI 越高屏幕画面越清晰。

（1）声明分辨率相关的图片资源

不同分辨率的图片，应该根据特定的目录结构来保存资源。结构如下所示：

- .../image.png
- .../Mx/image.png
- .../Nx/image.png

其中 M 和 N 是数字标识符，代表其中所包含图像的分辨率，也就是说，它们指定不同设备像素比例的图片。

主资源默认对应于 1.0 倍的分辨率图片。看一个例子：

- .../my_icon.png
- .../2.0x/my_icon.png
- .../3.0x/my_icon.png

在设备像素比率为 1.8 的设备上，将选择 .../2.0x/my_icon.png。对于 2.7 的设备像素比率，将选择 .../3.0x/my_icon.png。设备选择的是相近的图片资源。

（2）加载图片

要加载图片，请在 widget 的 build 方法中使用 AssetImage 类。例如，你的应用可以声明在 asset 中加载 logo。如下代码所示：

```
Widget build(BuildContext context) {

  return new DecoratedBox(
    decoration: new BoxDecoration(
      image: new DecorationImage(
        image: new AssetImage('images/logo.png'),
        // ...
```

```
        ),
        // ...
      ),
    );
    // ...
}
```

> 提示　使用默认的 Asset Bundle 加载资源时，内部会自动处理分辨率等，这些处理对开发者来说是无感知的。

3．依赖包中的资源图片

要加载依赖包中的图像，必须给 AssetImage 提供 package 参数。

例如，假设你的应用程序依赖于一个名为"resources"的包，它具有如下目录结构：

- .../pubspec.yaml
- .../assets/wifi.png
- .../assets /add.png
- .../ assets /phone.png

然后加载图像，使用 AssetImage，代码如下所示：

```
new AssetImage('assets/wifi.png', package:
'resources')
```

包使用本身的资源也应该加上 package 参数来获取。

9.1.3　平台 assets

也有时候可以直接在平台项目中使用 asset。以下是在 Flutter 框架加载并运行之前使用资源的两种常见情况。

1．更新应用图标

更新 Flutter 应用程序启动图标的方式与在本机 Android 或 iOS 应用程序中更新启动图标的方式相同，图 9-2 为示例工程的图标。

图 9-2　App 应用图标

> 提示　关于启动图标的添加方式参见第 15 章。

2．更新启动页

在 Flutter 框架加载时，Flutter 会使用本地平台机制绘制启动页。此启动页将持续到 Flutter 渲染应用程序的第一帧时。启动页效果如图 9-3 所示。

图 9-3　应用程序启动页

> **注意**　启动页将持续到 Flutter 渲染应用程序的第一帧时才消失。这意味着如果不在应用程序的 main() 方法中调用 runApp 函数（或者更具体地说，如果不调用 window.render 去响应 window.onDrawFrame）的话，启动屏幕将永远持续显示。

（1）Android

要将启动屏幕添加到 Flutter 应用程序，请导航至 drawable 目录，如图 9-4 所示。打开 launch_background.xml，通过自定义 drawable 来实现自定义启动界面。

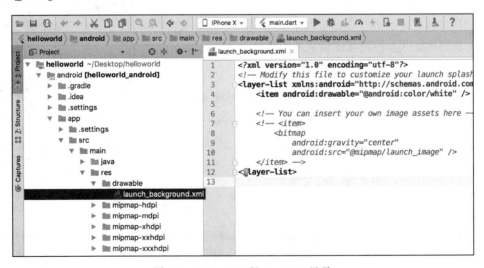

图 9-4　Android 工程 drawable 目录

（2）iOS

要将图片添加到启动屏幕（splash screen）的中心，请导航至 .../ios/Runner。在 Assets.xcassets/LaunchImage.imageset，如图 9-5 所示，拖入图片，并命名为 LaunchImage.png、LaunchImage@2x.png、LaunchImage@3x.png。

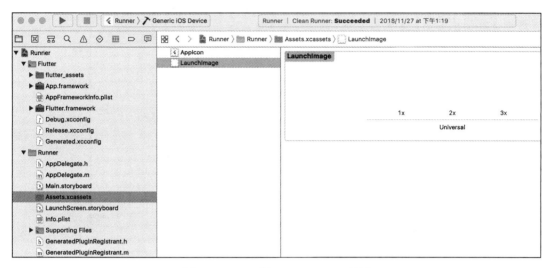

图 9-5　iOS 工程 Assets.xcassets 目录

9.2　自定义字体

操作系统所带的字体往往不能满足应用场景的需要，这时候就需要使用特殊字体处理。接下来编写一个自定义字体的综合小例子。具体步骤如下：

步骤 1：在 helloworld 工程下新建一个 fonts 文件夹，并放一个字体文件，如图 9-6 所示。

步骤 2：打开工程根目录下的工程配置文件，如图 9-7 所示。

图 9-6　添加字体文件

图 9-7　pubspec.yaml 工程配置文件

步骤 3：修改工程配置文件，添加自定义字体配置。其中 family 为字体名称，fonts 为字体路径，请一定要填写正确，否则会报错，如图 9-8 所示。

```
# To add custom fonts to your application, add a fonts section here,
# in this "flutter" section. Each entry in this list should have a
# "family" key with the font family name, and a "fonts" key with a
# list giving the asset and other descriptors for the font. For
fonts:
  - family: myfont
    fonts:
      - asset: fonts/myfont.ttf
# example:
# fonts:
#   - family: Schyler
#     fonts:
#       - asset: fonts/Schyler-Regular.ttf
#       - asset: fonts/Schyler-Italic.ttf
#         style: italic
#   - family: Trajan Pro
#     fonts:
#       - asset: fonts/TrajanPro.ttf
#       - asset: fonts/TrajanPro_Bold.ttf
#         weight: 700
#
# For details regarding fonts from package dependencies,
# see https://flutter.io/custom-fonts/#from-packages
```

图 9-8　自定义字体资源配置

步骤 4：编写自定义示例代码，完整的示例代码如下：

```dart
import 'package:flutter/material.dart';

void main() {
  runApp(new MaterialApp(
    title: '自定义字体',
    home: new MyApp(),
  ));
}


class MyApp extends StatelessWidget {


  @override
  Widget build(BuildContext context) {
    return new Scaffold(
      appBar: new AppBar(
        title: new Text('自定义字体'),
      ),
      body: new Center(
        child: new Text(
          '你好 flutter',
          style: new TextStyle(fontFamily: 'myfont',fontSize: 36.0),
        ),
```

```
        ),
      );
    }
}
```

上述示例代码的视图展现大致如图 9-9 所示。

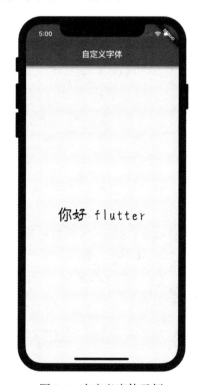

图 9-9　自定义字体示例

Chapter 10 第 10 章

路由及导航

前面章节讲解的基础组件和页面布局都是 UI 里相对基础的内容。这一章会给大家详细讲解 Flutter 页面的路由及导航是如何处理的。在本章里不仅仅会谈到页面的渲染,还会阐述数据是如何传递的。

本章围绕以下几个方面展开:
- 页面跳转基本使用
- 页面跳转发送数据
- 页面跳转返回数据

10.1 页面跳转基本使用

移动应用通常通过称为"屏幕"或"页面"的全屏元素显示其内容,在 Flutter 中,这些元素称为路由,它们由导航器 Navigator 组件管理。导航器管理一组路由 Route 对象,并提供了管理堆栈的方法,例如 Navigator.push 和 Navigator.pop。

接下来我们通过一个查看商品页面跳转的示例来讲解页面跳转的基本使用方法。具体步骤如下:

步骤 1:新建第一个页面叫 FirstScreen,添加一个"查看商品详情页面"的按钮,用来示范按下跳转处理。关键代码如下所示:

```
body: new Center(
  child: new RaisedButton(
    child: new Text('查看商品详情页面'),
    onPressed: (){
```

```
        Navigator.push(
          context,
          new MaterialPageRoute(builder: (context) => new SecondScreen())
        );
      },
    ),
  ),
),
```

FirstScreen 效果图如图 10-1 所示。

步骤 2：再添加一个页面命名为 SecondScreen，页面里添加一个"返回页面"按钮，按下则会返回到第一个页面，关键代码如下所示：

```
body: new Center(
    child: new RaisedButton(
    onPressed: (){
      Navigator.pop(context);
    },
    child: new Text('返回页面'),
  ),
),
```

SecondScreen 页面效果如图 10-2 所示。

图 10-1　查看商品详情页面操作

图 10-2　返回页面操作

完整示例代码如下所示：

```dart
import 'package:flutter/material.dart';

void main() {
  runApp(new MaterialApp(
    title: '导航页面示例',
    home: new FirstScreen(),
  ));
}


class FirstScreen extends StatelessWidget {


  @override
  Widget build(BuildContext context) {
    return new Scaffold(
      appBar: new AppBar(
        title: new Text('导航页面示例'),
      ),
      body: new Center(
        child: new RaisedButton(
          child: new Text('查看商品详情页面'),
          onPressed: (){
            Navigator.push(
                context,
                new MaterialPageRoute(builder: (context) => new SecondScreen())
            );
          },
        ),
      ),
    );
  }
}

class SecondScreen extends StatelessWidget {
  @override
  Widget build(BuildContext context) {

    return new Scaffold(
      appBar: new AppBar(
        title: new Text("导航页面示例"),
      ),
      body: new Center(
        child: new RaisedButton(
          onPressed: (){
            Navigator.pop(context);
          },
          child: new Text('返回页面'),
```

```
      ),
    ),
  );
 }
}
```

10.2 页面跳转发送数据

页面跳转时有时需要发送数据到第二个页面，比如从订单列表到商品详情页面时，通常需要发送商品 id 参数。接下来我们通过一个商品列表跳转到商品页面的示例来讲解页面跳转发送数据的实现方法。具体步骤如下：

步骤 1：新建第一个页面叫 ProductList，使用 List.generate 方法构造 20 条模拟数据。主要代码如下所示：

```
home: new ProductList(
  products:
  new List.generate(20, (i) => new Product('商品 $i', '这是一个商品详情 $i')),
),
```

接着再使用 ListView.builder 方法渲染商品列表，主要代码如下所示：

```
body: new ListView.builder(
  itemCount: products.length,
  itemBuilder: (context, index) {
    return new ListTile(
      title: new Text(products[index].title),
      onTap: () {
        Navigator.push(
          context,
          new MaterialPageRoute(
            builder: (context) =>
            new ProductDetail(product: products[index])),
        );
      },
    );
  }),
```

> **注意** 代码中 new ProductDetail(product: products[index])) 即为发送 product 数据。

ProductList 的效果图如图 10-3 所示。

步骤 2：当按下商品列表里某一项时，页面会跳转至商品详情页面，商品详情页面命名为 ProductDetail，此页面展示商品的标题及描述内容，主要代码如下所示：

```
@override
 Widget build(BuildContext context) {
```

```
    return new Scaffold(
      appBar: new AppBar(
        title: new Text("${product.title}"),
      ),
      body: new Padding(
        padding: new EdgeInsets.all(16.0),
        child: new Text('${product.description}'),
      ),
    );
  }
```

ProductDetail 的页面效果如图 10-4 所示。

图 10-3　商品列表页面　　　　　　图 10-4　商品详情页面

完整的示例代码如下所示：

```
import 'package:flutter/material.dart';

class Product {
  final String title;
  final String description;

  Product(this.title, this.description);
}
```

```dart
void main() {
  runApp(new MaterialApp(
    title: '传递数据示例',
    home: new ProductList(
      products:
      new List.generate(20, (i) => new Product('商品 $i', '这是一个商品详情 $i')),
    ),
  ));
}

class ProductList extends StatelessWidget {
  final List<Product> products;

  ProductList({Key key, @required this.products}) : super(key: key);

  @override
  Widget build(BuildContext context) {
    return new Scaffold(
      appBar: new AppBar(
        title: new Text("商品列表"),
      ),
      body: new ListView.builder(
          itemCount: products.length,
          itemBuilder: (context, index) {
            return new ListTile(
              title: new Text(products[index].title),
              onTap: () {
                Navigator.push(
                  context,
                  new MaterialPageRoute(
                      builder: (context) =>
                      new ProductDetail(product: products[index])),
                );
              },
            );
          }),
    );
  }
}

class ProductDetail extends StatelessWidget {
  final Product product;

  ProductDetail({Key key, @required this.product}) : super(key: key);

  @override
  Widget build(BuildContext context) {
    return new Scaffold(
      appBar: new AppBar(
        title: new Text("${product.title}"),
```

```
    ),
    body: new Padding(
      padding: new EdgeInsets.all(16.0),
      child: new Text('${product.description}'),
    ),
  );
 }
}
```

10.3 页面跳转返回数据

页面跳转不仅要发送数据，有时还需要从第二个页面返回数据，接下来我们通过一个示例来展示页面跳转返回数据的实现方法。具体步骤如下：

步骤 1：新建第一个页面叫 FirstPage，编写一个跳转至第二个页面的方法，主要代码如下所示：

```
_navigateToSecondPage(BuildContext context) async {

    final result = await Navigator.push(
      context,
      new MaterialPageRoute(builder: (context) => new SecondPage()),
    );

    Scaffold.of(context).showSnackBar(new SnackBar(content: new Text
      ("$result")));

}
```

接着 new 一个 RaisedButton 按钮，在它的 onPressed 方法里调用上面那个方法，代码如下：

```
return new RaisedButton(
  onPressed: (){
    _navigateToSecondPage(context);
  },
  child: new Text(' 跳转到第二个页面 '),
);
```

FirstPage 的效果图如图 10-5 所示。

步骤 2：当按下"跳转到第二个页面"时，页面会跳转至第二个页面，页面的类名命名为 SecondPage，在第二个页面里需要添加一个按钮，调用 Navigator.pop 方法来返回至第一个页面，主要代码如下所示：

```
child: new RaisedButton(
  onPressed: (){
```

```
        Navigator.pop(context,'hi flutter');
      },
      child: new Text('hi flutter'),
    ),
```

SecondPage 的页面效果如图 10-6 所示。

图 10-5　页面跳转返回数据页面一

图 10-6　页面跳转返回数据页面二

当按下"hi flutter"按钮后，页面又返回至第一个页面，如图 10-7 所示。同时页面的下方会弹出一个提示消息为"hi flutter"，表示返回数据成功。提示消息的代码如下：

```
Scaffold.of(context).showSnackBar(new SnackBar(content: new Text("$result")));
```

完整的示例代码如下所示：

```
import 'package:flutter/material.dart';




void main() {
  runApp(new MaterialApp(
    title: '页面跳转返回数据示例',
    home: new FirstPage(),
  ));
```

```dart
}


class FirstPage extends StatelessWidget {

  @override
  Widget build(BuildContext context) {
    return new Scaffold(
      appBar: new AppBar(
        title: new Text("页面跳转返回数据示例"),
      ),
      body: new Center(child: new RouteButton(),),
    );
  }
}

class RouteButton extends StatelessWidget {
  @override
  Widget build(BuildContext context) {

    return new RaisedButton(
      onPressed: (){
        _navigateToSecondPage(context);
      },
      child: new Text('跳转到第二个页面'),
    );
  }

  _navigateToSecondPage(BuildContext context) async {

    final result = await Navigator.push(
      context,
      new MaterialPageRoute(builder: (context) => new SecondPage()),
    );

    Scaffold.of(context).showSnackBar(new SnackBar(content: new Text
      ("$result")));

  }
}


class SecondPage extends StatelessWidget {

  @override
  Widget build(BuildContext context) {
    return new Scaffold(
```

图 10-7 页面跳转返回数据页面三

```
      appBar: new AppBar(
        title: new Text(" 选择一条数据 "),
      ),
      body: new Center(
        child: new Column(
          mainAxisAlignment: MainAxisAlignment.center,
          children: <Widget>[
            new Padding(
              padding: const EdgeInsets.all(8.0),
              child: new RaisedButton(
                onPressed: (){
                  Navigator.pop(context,'hi google');
                },
                child: new Text('hi google'),
              ),
            ),
            new Padding(
              padding: const EdgeInsets.all(8.0),
              child: new RaisedButton(
                onPressed: (){
                  Navigator.pop(context,'hi flutter');
                },
                child: new Text('hi flutter'),
              ),
            ),
          ],
        ),
      ),
    );
  }
}
```

第 11 章
组件装饰和视觉效果

类似于平时家里房子装饰，要想把界面做得漂亮和酷炫就需要一些特殊处理。比如添加边框、增加透明度、做一部分裁剪等。本章介绍几个常规的装饰和视图效果的处理方法。最后围绕着自定义画板这个主题讲解相关的组件使用方法。

本章涉及的组件及内容有：

- Opacity（透明度处理）
- DecoratedBox（装饰盒子）
- RotatedBox（旋转盒子）
- Clip（剪裁处理）
- 自定义画板案例

11.1 Opacity（透明度处理）

Opacity 组件里有一个 opacity 属性，能调整子组件的不透明度，使子控件部分透明，不透明度的量从 0.0 到 1.0 之间，0.0 表示完全透明，1.0 表示完全不透明。这样你就可以根据需要做出半透明或者毛玻璃的效果。很多应用做成半透明形式瞬间会显得高大上起来。比如：手机的相机大多都采用半透明效果。

Opacity 组件的主要属性如下所示：

属性名	类型	说明
opacity	double	不透明度值，不透明度的量从 0.0 到 1.0 之间，0.0 表示完全透明，1.0 表示完全不透明。
child	Widget	组件的子组件，只能有一个子组件，子组件受不透明度属性影响。

接下来编写一个例子，添加一个容器，外围放一个 Opacity 组件包装，不透明度值设置为 0.3，容器加了一个纯黑的底色，主要是为了演示不透明度值对它的显示影响程度，当设置了 0.3 以后你会发现容器变成了灰灰的颜色。

完整的示例代码如下：

```
import 'package:flutter/material.dart';

class LayoutDemo extends StatelessWidget {
  @override
  Widget build(BuildContext context) {
    return new Scaffold(
      appBar: new AppBar(
        title: new Text('Opacity 不透明度示例'),
      ),
      body: new Center(
        child: new Opacity(
          opacity: 0.3, // 不透明度设置为 0.3
          child: new Container(
            width: 250.0,
            height: 100.0,
            decoration: new BoxDecoration(
              color: Colors.black, // 背景色设置为纯黑
            ),
            child: Text(
              '不透明度为 0.3',
              style: TextStyle(
                color: Colors.white,
                fontSize: 28.0,
              ),
            ),
          ),
        ),
      ),
    );
  }
}

void main() {
  runApp(
    new MaterialApp(
      title: 'Opacity 不透明度示例',
      home: new LayoutDemo(),
    ),
  );
}
```

上述示例代码的视图展现大致如图 11-1 所示。

图 11-1　Opacity 组件示例

11.2 DecoratedBox（装饰盒子）

DecoratedBox 可以从多方面进行装饰处理。如颜色、形状、阴影、渐变及背景图片等。它有一个非常重要的属性是 decoration，类型为 BoxDecoration。

那么我们重点列一下 BoxDecoration 的主要属性，如表 11-1 所示。

表 11-1 BoxDecoration 组件属性及描述

属性名	类型	默认值	说明
shape	BoxShape	BoxShape.rectangle	形状取值
corlor	Corlor		用来渲染容器的背影色
boxShadow	List<BoxShadow>		阴影效果
gradient	Gradient		渐变色取值有：线性渐变、环形渐变
image	DecorationImage		背景图片
border	BoxBorder		边框样式
borderRadius	BorderRadiusGeometry		边框的弧度

接下来编写几个例子来演示各种装饰的效果。

1. 背景图效果

给容器添加背景图，只需要给 image 属性指定一个 DecorationImage 对象就行。它和 Image 的属性基本一致。

示例代码如下所示：

```
import 'package:flutter/material.dart';
class LayoutDemo extends StatelessWidget {
  @override
  Widget build(BuildContext context) {
    return new Scaffold(
      appBar: new AppBar(
        title: new Text('BoxDecoration 装饰盒子-背景图示例'),
      ),
      body: new Center(
        child: Container(
          width: 300.0,
          height: 300.0,
          decoration: BoxDecoration(
            color: Colors.grey,
            image: DecorationImage(
              image: ExactAssetImage('images/1.jpeg'),// 添加 image 属性
              fit: BoxFit.cover,// 填充类型
            ),
          ),
        )
      ),
    );
  }
}
```

```
}
void main() {
  runApp(
    new MaterialApp(
      title: 'BoxDecoration装饰盒子 – 背景图示例',
      home: new LayoutDemo(),
    ),
  );
}
```

上述示例代码的视图展现大致如图 11-2 所示。

2.边框圆角处理

给容器添加边框，既可以添加所有边框，也可以只加某一个边框。为了使得容器显得平滑，可以添加 borderRadius 属性值，值越大弧度越大。边框设置的代码如下所示：

```
border: Border.all(color: Colors.grey, width:
    4.0),
```

其中 EdgeInsets 支持多种自定义方法：

❏ EdgeInsets.all() 所有方向。
❏ EdgeInsets.only(left，top，right，bottom) 分别定义各个方向的边框。
❏ EdgeInsets.symmetric(vertical，horizontal) 自定义垂直，水平方向边框。
❏ EdgeInsets.fromWindowPadding(ui.WindowPadding padding，double devicePixelRatio) 根据机型屏幕尺寸定义。

图 11-2　背景图示例

在"背景图效果"的示例基础上，添加一个边框及圆角处理。完整的示例代码如下所示：

```
import 'package:flutter/material.dart';

class LayoutDemo extends StatelessWidget {
  @override
  Widget build(BuildContext context) {
    return new Scaffold(
      appBar: new AppBar(
        title: new Text('BoxDecoration装饰盒子 – 边框圆角示例'),
      ),
      body: new Center(
        child: Container(
          width: 260.0,
          height: 260.0,
```

```
            decoration: BoxDecoration(
              color: Colors.white,
              //添加所有的边框，颜色为灰色，宽度为4.0
              border: Border.all(color: Colors.grey, width: 4.0),
              //添加边框弧度，这样会有一个圆角效果
              borderRadius: BorderRadius.circular(36.0),
              image: DecorationImage(
                image: ExactAssetImage('images/1.jpeg'), //添加 image 属性
                fit: BoxFit.cover, //填充类型
              ),
            ),
          ),
        ),
      );
    }
  }

  void main() {
    runApp(
      new MaterialApp(
        title: 'BoxDecoration装饰盒子-边框圆角示例',
        home: new LayoutDemo(),
      ),
    );
  }
```

上述示例代码的视图展现大致如图 11-3 所示。

3. 边框阴影处理

为容器的边框加上阴影，会使得容器显得更有立体感。在 DecoratedBox 组里添加 boxShadow 即可实现。BoxShadow 有几个重要属性，如下所示：

- color：阴影颜色
- blurRadius：模糊值
- spreadRadius：扩展阴影半径
- offset：x 和 y 方向偏移量

BoxShadow 的使用如下代码所示：

```
boxShadow: <BoxShadow>[
  new BoxShadow(
    color: Colors.grey, //阴影颜色
    blurRadius: 8.0, //模糊值
    spreadRadius: 8.0, //扩展阴影半径
    offset: Offset(-1.0, 1.0), //x和y方向偏移量
  ),
],
```

图 11-3　边框圆角处理示例

编写一个例子，添加一个容器并加上 BoxShadow 处理。完整的示例代码如下所示：

```
import 'package:flutter/material.dart';

class LayoutDemo extends StatelessWidget {
  @override
  Widget build(BuildContext context) {
    return new Scaffold(
      appBar: new AppBar(
        title: new Text('BoxDecoration装饰盒子-边框阴影示例'),
      ),
      body: new Center(
        child: Container(
          width: 300.0,
          height: 300.0,
          decoration: BoxDecoration(
            color: Colors.white,
            //边框阴影效果
            boxShadow: <BoxShadow>[
              new BoxShadow(
                color: Colors.grey, //阴影颜色
                blurRadius: 8.0, //模糊值
                spreadRadius: 8.0, //扩展阴影半径
                offset: Offset(-1.0, 1.0), //x和y方向偏移量
              ),
            ],
          ),
          child: Text(
            'BoxShadow阴影效果',
            style: TextStyle(
              color: Colors.black,
              fontSize: 28.0,
            ),
          ),
        ),
      ),
    );
  }
}

void main() {
  runApp(
    new MaterialApp(
      title: 'BoxDecoration装饰盒子-边框阴影示例',
      home: new LayoutDemo(),
    ),
  );
}
```

上述示例代码的视图展现大致如图 11-4 所示。

图 11-4　边框阴影处理示例

4. 渐变处理

渐变有两种形式,一种是 LinearGradient 线性渐变,另一种是 RadialGradient 环形渐变。不管是哪种渐变形式都有一个共性,即需要一组数组数据来进行逐步渲染界面。

LinearGradient 线性渐变参数包括:

- begin:起始偏移量。
- end:终止偏移量。
- colors:渐变颜色数据集。

LinearGradient 的使用如下代码所示:

```
gradient: new LinearGradient(
  begin: const FractionalOffset(0.5, 0.0),// 起始偏移量
  end: const FractionalOffset(1.0, 1.0),// 终止偏移量
  // 渐变颜色数据集
  colors: <Color>[
    Colors.red,
    Colors.green,
    Colors.blue,
    Colors.grey,
  ],
),
```

编写一个例子,添加一个容器并加上线性渐变处理。完整的示例代码如下所示:

```
import 'package:flutter/material.dart';

class LayoutDemo extends StatelessWidget {
  @override
  Widget build(BuildContext context) {
    return new Scaffold(
      appBar: new AppBar(
        title: new Text('LinearGradient 线性渐变效果 '),
      ),
      body: new Center(
        child: new DecoratedBox(
          decoration: new BoxDecoration(
            gradient: new LinearGradient(
              begin: const FractionalOffset(0.5, 0.0),// 起始偏移量
              end: const FractionalOffset(1.0, 1.0),// 终止偏移量
              // 渐变颜色数据集
              colors: <Color>[
                Colors.red,
                Colors.green,
                Colors.blue,
                Colors.grey,
              ],
            ),
```

```
        ),
        child: new Container(
          width: 280.0,
          height: 280.0,
          child: new Center(
            child: Text(
              'LinearGradient 线性渐变效果',
              style: TextStyle(
                color: Colors.black,
                fontSize: 28.0,
              ),
            ),
          ),
        ),
      ),
    );
  }
}

void main() {
  runApp(
    new MaterialApp(
      title: 'DecoratedBox 装饰盒子示例',
      home: new LayoutDemo(),
    ),
  );
}
```

图 11-5　LinearGradient 线性渐变示例

上述示例代码的视图展现大致如图 11-5 所示。

RadialGradient 环形渐变参数包括：

❏ center：中心点偏移量，即 x 和 y 方向偏移量。

❏ radius：圆形半径。

❏ colors：渐变颜色数据集

RadialGradient 的使用如下代码所示：

```
gradient: RadialGradient(
  center: const Alignment(-0.0, -0.0), // 中心点偏移量,x 和 y 均为 0.0 表示在正中心位置
  radius: 0.50, // 圆形半径
  // 渐变颜色数据集
  colors: <Color>[
    Colors.red,
    Colors.green,
    Colors.blue,
    Colors.grey,
  ],
),
```

编写一个例子，添加一个容器并加上环形渐变处理。完整的示例代码如下所示：

```
import 'package:flutter/material.dart';

class LayoutDemo extends StatelessWidget {
  @override
  Widget build(BuildContext context) {
    return new Scaffold(
      appBar: new AppBar(
        title: new Text('RadialGradient 环形渐变效果'),
      ),
      body: new Center(
        child: new DecoratedBox(
          decoration: new BoxDecoration(
            gradient: RadialGradient(
              center: const Alignment(-0.0, -0.0), // 中心点偏移量,x和y均为0.0表示
                                                   // 在正中心位置
              radius: 0.50, // 圆形半径
              // 渐变颜色数据集
              colors: <Color>[
                Colors.red,
                Colors.green,
                Colors.blue,
                Colors.grey,
              ],
            ),
          ),
          child: new Container(
            width: 280.0,
            height: 280.0,
            child: new Center(
              child: Text(
                'RadialGradient 环形渐变效果',
                style: TextStyle(
                  color: Colors.black,
                  fontSize: 28.0,
                ),
              ),
            ),
          ),
        ),
      ),
    );
  }
}

void main() {
  runApp(
    new MaterialApp(
      title: 'DecoratedBox 装饰盒子示例',
      home: new LayoutDemo(),
    ),
  );
}
```

上述示例代码的视图展现大致如图 11-6 所示。

图 11-6　RadialGradient 环形渐变示例

11.3 RotatedBox（旋转盒子）

RotatedBox 组件即为旋转组件，可以使得 child 发生旋转，旋转的度数是 90 度的整数倍。每一次旋转只能是 90 度。例如当它的属性 quarterTurns 为 3 时，表示旋转了 270 度。旋转盒子通常用于图片的旋转。比如在相册里，用户想把照片横着看或者竖着看，那么旋转盒子使用起来就非常方便了。

接下来我们写一个例子，添加一个文本，让它旋转 3 次，即旋转 270 度。完整的示例代码如下：

```
import 'package:flutter/material.dart';

void main() => runApp(new MyApp());

class MyApp extends StatelessWidget {
  @override
  Widget build(BuildContext context) {
    return new MaterialApp(
      title: 'RotatedBox旋转盒子示例',
      home: Scaffold(
        appBar: AppBar(
          title: Text(
            'RotatedBox旋转盒子示例',
            style: TextStyle(color: Colors.white),
          ),
        ),
        body: Center(
          child: RotatedBox(
            quarterTurns: 3,// 旋转次数，一次为90度
            child: Text(
              'RotatedBox旋转盒子',
              style: TextStyle(fontSize: 28.0),
            ),
          ),
        ),
      ),
    );
  }
}
```

上述示例代码的视图展现大致如图 11-7 所示。

11.4 Clip（剪裁处理）

Clip 功能是把一个组件剪掉一部分。Flutter 里有多个组件可以完成此类功能，如下所示：

❏ ClipOval：圆形剪裁。
❏ ClipRRect：圆角矩形剪裁。

图 11-7 RotatedBox 组件示例

- ClipRect：矩形剪裁。
- ClipPath：路径剪裁。

这几类剪裁组件都有两个共同属性如下所示：

属性名	类型	说明
clipper	CustomClipper<Path>	剪裁路径，比如椭圆、矩形等。
clipBehavior	Clip	剪裁方式。

1. ClipOval 圆形剪裁

圆形剪裁可以用来剪裁圆形头像，做一个类似 Avatar 的组件。示例代码如下所示：

```
import 'package:flutter/material.dart';

void main() => runApp(new MyApp());

class MyApp extends StatelessWidget {
  @override
  Widget build(BuildContext context) {
    return new MaterialApp(
      title: 'ClipOval 圆形剪裁示例',
      home: Scaffold(
        appBar: AppBar(
          title: Text(
            'ClipOval 圆形剪裁示例',
            style: TextStyle(color: Colors.white),
          ),
        ),
        body: Center(
          child: new ClipOval(
            child: new SizedBox(
              width: 300.0,
              height: 300.0,
              child: new Image.asset(
                "images/8.jpeg",
                fit: BoxFit.fill,
              ),
            ),
          ),
        ),
      ),
    );
  }
}
```

上述示例代码的视图展现大致如图 11-8 所示。

2. ClipRRect 圆角矩形剪裁

ClipRRect 这个组件用 borderRadius 参数来控制圆角的位置大小。示例代码如下所示：

图 11-8　ClipOval 组件示例

```
import 'package:flutter/material.dart';

void main() => runApp(new MyApp());

class MyApp extends StatelessWidget {
  @override
  Widget build(BuildContext context) {
    return new MaterialApp(
      title: 'ClipRRect圆角矩形剪裁示例',
      home: Scaffold(
        appBar: AppBar(
          title: Text(
            'ClipRRect圆角矩形剪裁示例',
            style: TextStyle(color: Colors.white),
          ),
        ),
        body: Center(
          child: new ClipRRect(
            borderRadius: new BorderRadius.all(
                new Radius.circular(30.0)),// 圆角弧度,值越大弧度越大
            child: new SizedBox(
              width: 300.0,
              height: 300.0,
              child: new Image.asset(
                "images/8.jpeg",
                fit: BoxFit.fill,
              ),
            ),
          ),
        ),
      ),
    );
  }
}
```

上述示例代码的视图展现大致如图 11-9 所示。

3. ClipRect 矩形剪裁

ClipRect 这个组件需要自定义 clipper 属性才能使用，否则没有效果。自定义 clipper 需要继承 CustomClipper 类，并且需要重写 getClip 及 shouldReclip 两个方法。

完整的示例代码如下所示：

```
import 'package:flutter/material.dart';

void main() => runApp(new MyApp());

class MyApp extends StatelessWidget {
```

图 11-9　ClipRRect 组件示例

```
  @override
  Widget build(BuildContext context) {
    return new MaterialApp(
      title: 'ClipRect 矩形剪裁示例',
      home: Scaffold(
        appBar: AppBar(
          title: Text(
            'ClipRect 矩形剪裁示例',
            style: TextStyle(color: Colors.white),
          ),
        ),
        body: Center(
          child: new ClipRect(
            clipper: new RectClipper(),// 指定自定义 clipper
            child:new SizedBox(
              width: 300.0,
              height:300.0,
              child:  new Image.asset("images/8.jpeg",fit: BoxFit.fill,),
            ) ,
          ),
        ),
      ),
    );
  }
}

// 自定义 clipper，类型为 Rect
class RectClipper extends CustomClipper<Rect>{

  // 重写获取剪裁范围
  @override
  Rect getClip(Size size) {
    return new Rect.fromLTRB(100.0, 100.0, size.
      width - 100.0,  size.height- 100.0);
  }

  // 重写是否重新剪裁
  @override
  bool shouldReclip(CustomClipper<Rect>
    oldClipper) {
    return true;
  }

}
```

上述示例代码的视图展现大致如图 11-10 所示。

4. ClipPath 路径剪裁

ClipRect 这个组件的想像空间就大了，由于采用了矢

图 11-10　ClipRect 组件示例

量路径 path，所以可以把组件剪裁为任意类型的形状。比如三角形、矩形、星形及多边形等等。自定义 clipper 需要继承 CustomClipper 类，并且需要重写 getClip 及 shouldReclip 两个方法。

完整的示例代码如下所示：

```
import 'package:flutter/material.dart';
import 'dart:math';

void main() => runApp(new MyApp());

class MyApp extends StatelessWidget {
  @override
  Widget build(BuildContext context) {
    return new MaterialApp(
      title: 'ClipPath 路径剪裁示例',
      home: Scaffold(
        appBar: AppBar(
          title: Text(
            'ClipPath 路径剪裁示例',
            style: TextStyle(color: Colors.white),
          ),
        ),
        body: Center(
          child: new ClipPath(
            clipper: new TriangleCliper(),// 指定自定义三角形 clipper
            child:new SizedBox(
              width: 100.0,
              height:100.0,
              child:  new Image.asset("images/8.jpeg",fit: BoxFit.fill,),
            ),
          ),
        ),
      ),
    );
  }
}


// 自定义三角形 clipper, 类型为 Path
class TriangleCliper extends CustomClipper<Path>{

  // 重写获取剪裁范围
  @override
  Path getClip(Size size) {

    Path path = new Path();
    path.moveTo(50.0,50.0);// 起始点
    path.lineTo(50.0,10.0);// 终止点
    path.lineTo(100.0,50.0);// 起始点 (50,10), 终止点
```

```
      path.close();// 使这些点构成三角形


      return path;
   }

   // 重写是否重新剪裁
   @override
   bool shouldReclip(TriangleCliper oldClipper) {
      return true;
   }

}
```

上述示例代码的视图展现大致如图 11-11 所示。

11.5 案例——自定义画板

案例——自定义画板可以画任意图形，如：点、线、路径、矩形、圆形以及添加图像。最典型的应用场景就是远程教育和视频会议里的电子白板功能。如图 11-12 所示是笔者实现的视频会议产品里的电子白板功能。电子白板里功能丰富，有画板、画笔、颜色粗细属性及添加图片等功能。接下来分别讲述一下这些概念。

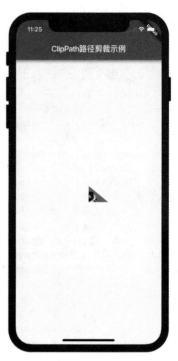

图 11-11　ClipPath 组件示例

图 11-12　电子白板

1. 画布 Canvas 与画笔 Paint

画布好比是教室里的黑板或者白板，画布是一个矩形区域，我们可以在上面任意涂鸦。在画布 Canvas 上可以画点、线、路径、矩形、圆形、以及添加图像。与画布 Canvas 相关的方法例如：

- 画直线：drawLine()
- 画圆：drawCircle ()
- 画椭圆：drawOval ()
- 画矩形：drawRect()
- 画点：drawPoints()
- 画圆弧：drawArc()

仅有画布还不行，还需要画笔，画笔 Paint 为绘制方法提供颜色及粗细等参数。如何创建画笔呢？需要提供一系列参数给画笔 Paint，如表 11-2 所示。

表 11-2　Paint 类参数说明

属性名	类型	参考值	说明
color	Colors	Colors.blueAccent	画笔颜色
strokeCap	StrokeCap	StrokeCap.round	画笔笔触类型
isAntiAlias	bool	true	是否启动抗锯齿
blendMode	BlendMode	BlendMode.exclusion	颜色混合模式
style	PaintingStyle	PaintingStyle.fill	绘画样式，默认为填充
colorFilter	ColorFilter	ColorFilter.mode (Colors.blueAccent, BlendMode.exclusion)	颜色渲染模式，一般是矩阵效果来改变的，但是 Flutter 中只能使用颜色混合模式
maskFilter	MaskFilter	MaskFilter.blur(BlurStyle.inner, 3.0)	模糊遮罩效果，Flutter 中只有这个
filterQuality	FilterQuality	FilterQuality.high	颜色渲染模式的质量
strokeWidth	double	16.0	画笔的粗细

 提示　颜色混合模式有很多种，详细请参考：https://docs.flutter.io/flutter/dart-ui/BlendMode-class.html

读懂这些参数后，实例化一支画笔即可，如下代码所示：

```
Paint _paint = Paint()
  ..color = Colors.green // 画笔颜色
  ..strokeCap = StrokeCap.round // 画笔笔触类型
  ..isAntiAlias = true // 是否启动抗锯齿
  ..blendMode = BlendMode.exclusion // 颜色混合模式
  ..style = PaintingStyle.fill // 绘画风格，默认为填充
  ..colorFilter = ColorFilter.mode(Colors.blueAccent,
      BlendMode.exclusion) // 颜色渲染模式
  ..maskFilter =
  MaskFilter.blur(BlurStyle.inner, 3.0) // 模糊遮罩效果
  ..filterQuality = FilterQuality.high // 颜色渲染模式的质量

  ..strokeWidth = 16.0 ;// 画笔的宽度
```

实际使用中不需要传入这么多参数,一般传入画笔颜色、粗细及填充色即可。

2. 绘制直线

绘制直线需要调用 Canvas 的 drawLine 方法,传入起点及终点的坐标即可。如下代码所示:

```
canvas.drawLine(Offset(20.0, 20.0), Offset(300.0, 20.0), _paint);
```

完整代码如下所示:

```
import 'package:flutter/material.dart';
import 'dart:math';

void main() => runApp(new MyApp());

class MyApp extends StatelessWidget {
  @override
  Widget build(BuildContext context) {
    return new MaterialApp(
      title: 'CustomPaint 绘制直线示例',
      home: Scaffold(
        appBar: AppBar(
          title: Text(
            'CustomPaint 绘制直线示例',
            style: TextStyle(color: Colors.white),
          ),
        ),
        body: Center(

          child: SizedBox(
            width: 500.0,
            height: 500.0,
            child: CustomPaint(
              painter: LinePainter(),
              child: Center(
                child: Text(
                  '绘制直线',
                  style: const TextStyle(
                    fontSize: 38.0,
                    fontWeight: FontWeight.w600,
                    color: Colors.black,
                  ),
                ),
              ),
            )
          ),
        ),
```

```
    );
  }
}
```

```
// 继承自 CustomPainter 并且实现 CustomPainter 里面的 paint 和 shouldRepaint 方法
class LinePainter extends CustomPainter {
  //定义画笔
  Paint _paint = new Paint()
    ..color = Colors.black
    ..strokeCap = StrokeCap.square
    ..isAntiAlias = true
    ..strokeWidth = 3.0
    ..style = PaintingStyle.stroke;

  ///重写绘制内容方法
  @override
  void paint(Canvas canvas, Size size) {
    //绘制直线
    canvas.drawLine(Offset(20.0, 20.0), Offset(300.0, 20.0), _paint);
  }

  ///重写是否需要重绘的
  @override
  bool shouldRepaint(CustomPainter oldDelegate) {
    return false;
  }
}
```

上述示例代码的视图展现大致如图11-13所示。

3. 绘制圆

绘制圆需要调用 Canvas 的 drawCircle 方法，需要传入中心点的坐标、半径及画笔即可。如下代码所示：

```
canvas.drawCircle(Offset(200.0, 150.0), 150.0, _paint);
```

其中画笔可以对应有填充色及没有填充色两种情况：
- PaintingStyle.fill：填充绘制。
- PaintingStyle.stroke：非填充绘制。

完整代码如下所示：

```
import 'package:flutter/material.dart';
import 'dart:math';

void main() => runApp(new MyApp());
```

图 11-13 CustomPaint 绘制直线示例

```dart
class MyApp extends StatelessWidget {
  @override
  Widget build(BuildContext context) {
    return new MaterialApp(
      title: 'CustomPaint 绘制圆示例',
      home: Scaffold(
        appBar: AppBar(
          title: Text(
            'CustomPaint 绘制圆示例',
            style: TextStyle(color: Colors.white),
          ),
        ),
        body: Center(

          child: SizedBox(
            width: 500.0,
            height: 500.0,
            child: CustomPaint(
              painter: LinePainter(),
              child: Center(
                child: Text(
                  '绘制圆',
                  style: const TextStyle(
                    fontSize: 38.0,
                    fontWeight: FontWeight.w600,
                    color: Colors.black,
                  ),
                ),
              ),
            ),
          )
        ),
      ),
    );
  }
}

// 继承自 CustomPainter 并且实现 CustomPainter 里面的 paint 和 shouldRepaint 方法
class LinePainter extends CustomPainter {

  // 定义画笔
  Paint _paint = new Paint()
    ..color = Colors.grey
    ..strokeCap = StrokeCap.square
    ..isAntiAlias = true
    ..strokeWidth = 3.0
    ..style = PaintingStyle.stroke;// 画笔样式有填充 PaintingStyle.fill 及没有填充
                                   //  PaintingStyle.stroke 两种

  // 重写绘制内容方法
```

```
@override
void paint(Canvas canvas, Size size) {
  // 绘制圆 参数为中心点，半径，画笔
  canvas.drawCircle(Offset(200.0, 150.0), 150.0, _paint);
}

// 重写是否需要重绘的
@override
bool shouldRepaint(CustomPainter oldDelegate) {
  return false;
}
}
```

上述示例代码的视图展现大致如图 11-14 所示。图中展示了填充及非填充两种样式。

图 11-14　CustomPaint 绘制圆填充及非填充示例

4. 绘制椭圆

绘制椭圆需要调用 Canvas 的 drawOval 方法，同时需要使用一个矩形来确定绘制的范围，椭圆是在这个矩形之中内切的，其中第一个参数为左上角坐标，第二个参数为右下角坐标。如下代码所示：

```
Rect rect = Rect.fromPoints(Offset(80.0, 200.0), Offset(300.0, 300.0));
canvas.drawOval(rect, _paint);
```

另外画笔也是有填充色及没有填充色两种情况。具体参考上节"绘制圆"所述。

创建 Rect 有多种方式:

- fromPoints(Offset a, Offset b): 使用左上和右下角坐标来确定内切矩形的大小和位置。
- fromCircle({ Offset center, double radius }): 使用圆的中心点坐标和半径确定外切矩形的大小和位置。
- fromLTRB(double left, double top, double right, double bottom): 使用矩形的上下左右的 X、Y 边界值来确定矩形的大小和位置。
- fromLTWH(double left, double top, double width, double height): 使用矩形左上角的 X、Y 坐标及矩形的宽高来确定矩形的大小和位置。

完整代码如下所示:

```
import 'package:flutter/material.dart';
import 'dart:math';

void main() => runApp(new MyApp());

class MyApp extends StatelessWidget {
  @override
  Widget build(BuildContext context) {
    return new MaterialApp(
      title: 'CustomPaint 绘制椭圆示例',
      home: Scaffold(
        appBar: AppBar(
          title: Text(
            'CustomPaint 绘制椭圆示例',
            style: TextStyle(color: Colors.white),
          ),
        ),
        body: Center(
          child: SizedBox(
            width: 500.0,
            height: 500.0,
            child: CustomPaint(
              painter: LinePainter(),
              child: Center(
                child: Text(
                  '绘制椭圆',
                  style: const TextStyle(
                    fontSize: 38.0,
                    fontWeight: FontWeight.w600,
                    color: Colors.black,
                  ),
                ),
              ),
            ),
          )
        ),
```

```
      ),
    );
  }
}

// 继承自 CustomPainter 并且实现 CustomPainter 里面的 paint 和 shouldRepaint 方法
class LinePainter extends CustomPainter {

  // 定义画笔
  Paint _paint = new Paint()
    ..color = Colors.grey
    ..strokeCap = StrokeCap.square
    ..isAntiAlias = true
    ..strokeWidth = 3.0
    ..style = PaintingStyle.fill;// 画笔样式有填充 PaintingStyle.fill 及没有填充
      PaintingStyle.stroke 两种

  /// 重写绘制内容方法
  @override
  void paint(Canvas canvas, Size size) {
    // 绘制椭圆
    // 使用一个矩形来确定绘制的范围，椭圆是在这个矩形之中内切的，第一个参数为左上角坐标，第二个
      参数为右下角坐标
    Rect rect = Rect.fromPoints(Offset(80.0, 200.0), Offset(300.0, 300.0));
    canvas.drawOval(rect, _paint);
  }

  /// 重写是否需要重绘的
  @override
  bool shouldRepaint(CustomPainter oldDelegate) {
    return false;
  }
}
```

上述示例代码的视图展现大致如图 11-15 所示。图中展示了填充及非填充两种样式。

5. 绘制圆角矩形

绘制圆角矩形需要调用 Canvas 的 drawRRect 方法。另外画笔同样是有填充色及没有填充色两种情况。下面是一个范例。

首先创建一个中心点坐标为（200，200）边长为 100 的矩形，如下面代码所示：

```
Rect rect = Rect.fromCircle(center: Offset(200.0, 200.0), radius: 100.0);
```

然后再把这个矩形转换成圆形矩形，角度为 20。如下面代码所示：

```
RRect rrect = RRect.fromRectAndRadius(rect, Radius.circular(20.0));
```

最后绘制这个带圆角的矩形即可，如下面代码所示：

```
canvas.drawRRect(rrect, _paint);
```

图 11-15　CustomPaint 绘制椭圆填充及非填充示例

完整代码如下所示：

```
import 'package:flutter/material.dart';
import 'dart:math';

void main() => runApp(new MyApp());

class MyApp extends StatelessWidget {
  @override
  Widget build(BuildContext context) {
    return new MaterialApp(
      title: 'CustomPaint绘制圆角矩形示例',
      home: Scaffold(
        appBar: AppBar(
          title: Text(
            'CustomPaint绘制圆角矩形示例',
            style: TextStyle(color: Colors.white),
          ),
        ),
        body: Center(

          child: SizedBox(
            width: 500.0,
            height: 500.0,
            child: CustomPaint(
```

```
              painter: LinePainter(),
              child: Center(
                child: Text(
                  '绘制圆角矩形',
                  style: const TextStyle(
                    fontSize: 18.0,
                    fontWeight: FontWeight.w600,
                    color: Colors.black,
                  ),
                ),
              ),
            )
          ),
        ),
      );
    }
}

// 继承自 CustomPainter 并且实现 CustomPainter 里面的 paint 和 shouldRepaint 方法
class LinePainter extends CustomPainter {

  //定义画笔
  Paint _paint = new Paint()
    ..color = Colors.grey
    ..strokeCap = StrokeCap.square
    ..isAntiAlias = true
    ..strokeWidth = 3.0
    ..style = PaintingStyle.stroke;//画笔样式有填充 PaintingStyle.fill 及没有填充
      PaintingStyle.stroke 两种

  ///重写绘制内容方法
  @override
  void paint(Canvas canvas, Size size) {

    // 中心点坐标为 200,200 边长为 100
    Rect rect = Rect.fromCircle(center: Offset(200.0, 200.0), radius: 100.0);
    // 根据矩形创建一个角度为 20 的圆角矩形
    RRect rrect = RRect.fromRectAndRadius(rect, Radius.circular(20.0));
    // 开始绘制圆角矩形
    canvas.drawRRect(rrect, _paint);
  }

  ///是否需要重绘
  @override
  bool shouldRepaint(CustomPainter oldDelegate) {
    return false;
  }
}
```

上述示例代码的视图展现大致如图 11-16 所示。图中展示了填充及非填充两种样式。

图 11-16　CustomPaint 绘制圆角矩形填充及非填充示例

6. 绘制嵌套矩形

绘制嵌套矩形需要调用 Canvas 的 drawDRRect 方法。函数中的这个 D 就是 Double 的意思，也就是可以绘制两个矩形的。另外画笔同样也是有填充色及没有填充色两种情况。下面是一个范例。

首先初始化两个矩形，如下面代码所示：

```
Rect rect1 = Rect.fromCircle(center: Offset(150.0, 150.0), radius: 80.0);
Rect rect2 = Rect.fromCircle(center: Offset(150.0, 150.0), radius: 40.0);
```

然后再把这两个矩形转化成圆角矩形。如下面代码所示：

```
RRect outer = RRect.fromRectAndRadius(rect1, Radius.circular(20.0));
RRect inner = RRect.fromRectAndRadius(rect2, Radius.circular(10.0));
```

最后绘制这个嵌套矩形即可，参数传入外部矩形、内部矩形及画笔。如下面代码所示：

```
canvas.drawDRRect(outer, inner, _paint);
```

完整代码如下所示：

```
import 'package:flutter/material.dart';
```

```dart
import 'dart:math';

void main() => runApp(new MyApp());

class MyApp extends StatelessWidget {
  @override
  Widget build(BuildContext context) {
    return new MaterialApp(
      title: 'CustomPaint 绘制嵌套矩形示例',
      home: Scaffold(
        appBar: AppBar(
          title: Text(
            'CustomPaint 绘制嵌套矩形示例',
            style: TextStyle(color: Colors.white),
          ),
        ),
        body: Center(
          child: SizedBox(
            width: 500.0,
            height: 500.0,
            child: CustomPaint(
              painter: LinePainter(),
            ),
          )
        ),
      ),
    );
  }
}

// 继承自 CustomPainter 并且实现 CustomPainter 里面的 paint 和 shouldRepaint 方法
class LinePainter extends CustomPainter {

  // 定义画笔
  Paint _paint = new Paint()
    ..color = Colors.grey
    ..strokeCap = StrokeCap.square
    ..isAntiAlias = true
    ..strokeWidth = 3.0
    ..style = PaintingStyle.stroke;// 画笔样式有填充 PaintingStyle.fill 及没有填充
      PaintingStyle.stroke 两种

  /// 重写绘制内容方法
  @override
  void paint(Canvas canvas, Size size) {

    // 初始化两个矩形
    Rect rect1 = Rect.fromCircle(center: Offset(150.0, 150.0), radius: 80.0);
    Rect rect2 = Rect.fromCircle(center: Offset(150.0, 150.0), radius: 40.0);
    // 再把这两个矩形转化成圆角矩形
```

```
    RRect outer = RRect.fromRectAndRadius(rect1, Radius.circular(20.0));
    RRect inner = RRect.fromRectAndRadius(rect2, Radius.circular(10.0));
    canvas.drawDRRect(outer, inner, _paint);
  }

  ///是否需要重绘
  @override
  bool shouldRepaint(CustomPainter oldDelegate) {
    return false;
  }
}
```

上述示例代码的视图展现大致如图 11-17 所示。图中展示了填充及非填充两种样式。

图 11-17　CustomPaint 绘制嵌套矩形填充及非填充示例

> **提示**　嵌套矩形如果是填充样式，不会全部填充，中心位置会留空。

7. 绘制多个点

绘制多个点需要调用 Canvas 的 drawPoints 方法。传入的参数 PointMode 的枚举类型有三个，points（点）、lines（隔点连接线）及 polygon（相邻连接线）。这里以绘制多个点为例来讲解 drawPoints 的使用方法。如下面代码所示：

```
PointMode.points,
  [
```

```
        Offset(50.0, 60.0),
        Offset(40.0, 90.0),
        Offset(100.0, 100.0),
        Offset(300.0, 350.0),
        Offset(400.0, 80.0),
        Offset(200.0, 200.0),
      ],
      _paint..color = Colors.grey);
```

这里编写一个示例，示例中 strokeWidth 值设置为 20.0，目的是为了画笔更粗一些，这样能显示得更明显。画笔的 strokeCap 属性 StrokeCap.round 为圆点，StrokeCap.square 为方形，对应提供两种画笔笔触样式。完整代码如下所示：

```
import 'package:flutter/material.dart';
import 'dart:ui';

void main() => runApp(new MyApp());

class MyApp extends StatelessWidget {
  @override
  Widget build(BuildContext context) {
    return new MaterialApp(
      title: 'CustomPaint 绘制多个点示例',
      home: Scaffold(
        appBar: AppBar(
          title: Text(
            'CustomPaint 绘制多个点示例',
            style: TextStyle(color: Colors.white),
          ),
        ),
        body: Center(
            child: SizedBox(
          width: 500.0,
          height: 500.0,
          child: CustomPaint(
            painter: LinePainter(),
          ),
        )),
      ),
    );
  }
}

// 继承自 CustomPainter 并且实现 CustomPainter 里面的 paint 和 shouldRepaint 方法
class LinePainter extends CustomPainter {
  // 定义画笔
  Paint _paint = new Paint()
    ..color = Colors.grey
    ..strokeCap = StrokeCap.round//StrokeCap.round 为圆点 StrokeCap.square 为方形
    ..isAntiAlias = true
    ..strokeWidth = 20.0// 画笔粗细，值调大点，这样点看起来明显一些
    ..style = PaintingStyle.fill; // 用于绘制点时 PaintingStyle 值无效
```

```
/// 重写绘制内容方法
@override
void paint(Canvas canvas, Size size) {
  // 绘制点
  canvas.drawPoints(
      ///PointMode 的枚举类型有三个，points 点、lines 隔点连接线、polygon 相邻连接线
      PointMode.points,
      [
        Offset(50.0, 60.0),
        Offset(40.0, 90.0),
        Offset(100.0, 100.0),
        Offset(300.0, 350.0),
        Offset(400.0, 80.0),
        Offset(200.0, 200.0),
      ],
      _paint..color = Colors.grey);
}

/// 是否需要重绘
@override
bool shouldRepaint(CustomPainter oldDelegate) {
  return false;
}
}
```

上述示例代码的视图展现大致如图 11-18 所示。

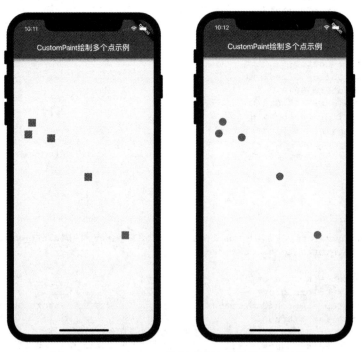

图 11-18　CustomPaint 绘制多个点方格及圆点示例

若更改上述示例代码中的 PointMode.points 为 PointMode.lines，则图像变为隔点连接线样式。视图展现大致如图 11-19 所示。

继续更改上述示例代码中的 PointMode.points 为 PointMode.polygon，则图像变为相邻连接线样式。视图展现大致如图 11-20 所示。

图 11-19　CustomPaint 绘制多个点
PointMode.lines 样式图

图 11-20　CustomPaint 绘制多个点
PointMode.polygon 样式图

8. 绘制圆弧

绘制圆弧需要调用 Canvas 的 drawArc 方法。需要传入绘制区域、弧度及画笔等参数。如下面代码所示：

```
// 定义矩形
Rect rect1 = Rect.fromCircle(center: Offset(100.0, 0.0), radius: 100.0);
// 画 1/2PI 弧度的圆弧
canvas.drawArc(rect1, 0.0, PI / 2, true, _paint);
```

这里写一个示例，需要用到一些几何知识。圆的整个弧度为 2π，示例中演示了 $\frac{1}{2}\pi$ 及 π 弧度的圆弧，也就是 1/4 及 1/2 个圆。完整代码如下所示：

```
import 'package:flutter/material.dart';
import 'dart:ui';
```

```dart
void main() => runApp(new MyApp());

class MyApp extends StatelessWidget {
  @override
  Widget build(BuildContext context) {
    return new MaterialApp(
      title: 'CustomPaint 绘制圆弧示例',
      home: Scaffold(
        appBar: AppBar(
          title: Text(
            'CustomPaint 绘制圆弧示例',
            style: TextStyle(color: Colors.white),
          ),
        ),
        body: Center(
            child: SizedBox(
          width: 500.0,
          height: 500.0,
          child: CustomPaint(
            painter: LinePainter(),
          ),
        )),
      ),
    );
  }
}

// 继承自 CustomPainter 并且实现 CustomPainter 里面的 paint 和 shouldRepaint 方法
class LinePainter extends CustomPainter {
  // 定义画笔
  Paint _paint = new Paint()
    ..color = Colors.grey
    ..strokeCap = StrokeCap.round
    ..isAntiAlias = true
    ..strokeWidth = 2.0 // 画笔粗细
    ..style = PaintingStyle.stroke; // 用于绘制点时 PaintingStyle 值无效

  /// 重写绘制内容方法
  @override
  void paint(Canvas canvas, Size size) {
    // 绘制圆弧
    const PI = 3.1415926;
    // 定义矩形
    Rect rect1 = Rect.fromCircle(center: Offset(100.0, 0.0), radius: 100.0);
    // 画 1/2PI 弧度的圆弧
    canvas.drawArc(rect1, 0.0, PI / 2, true, _paint);
    // 画 PI 弧度的圆弧
    Rect rect2 = Rect.fromCircle(center: Offset(200.0, 150.0), radius: 100.0);
    canvas.drawArc(rect2, 0.0, PI, true, _paint);
  }
```

```
/// 是否需要重绘
@override
bool shouldRepaint(CustomPainter oldDelegate) {
  return false;
}
}
```

上述示例代码的视图展现大致如图 11-21 所示。

9. 绘制路径 Path

使用 Canvas 的 drawPath 方法理论上可以绘制任意矢量图。Path 的主要方法如下所示：

❑ moveTo：将路径起始点移动到指定的位置。
❑ lineTo：从当前位置连接指定点。
❑ arcTo：曲线。
❑ conicTo：贝济埃曲线。
❑ close：关闭路径，连接路径的起始点。

这里写一个示例，调用 Canvas 的 drawPath 方法绘制一个任意图形，完整代码如下所示：

```
import 'package:flutter/material.dart';
import 'dart:ui';

void main() => runApp(new MyApp());

class MyApp extends StatelessWidget {
  @override
  Widget build(BuildContext context) {
    return new MaterialApp(
      title: 'CustomPaint 绘制路径示例',
      home: Scaffold(
        appBar: AppBar(
          title: Text(
            'CustomPaint 绘制路径示例',
            style: TextStyle(color: Colors.white),
          ),
        ),
        body: Center(
            child: SizedBox(
          width: 500.0,
          height: 500.0,
          child: CustomPaint(
            painter: LinePainter(),
          ),
        )),
      ),
```

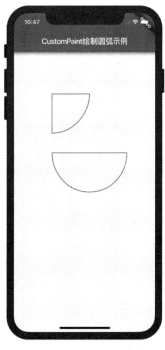

图 11-21 CustomPaint 绘制圆弧示例

```
      );
    }
}

// 继承自 CustomPainter 并且实现 CustomPainter 里面的 paint 和 shouldRepaint 方法
class LinePainter extends CustomPainter {
  // 定义画笔
  Paint _paint = new Paint()
    ..color = Colors.grey
    ..strokeCap = StrokeCap.round
    ..isAntiAlias = true
    ..strokeWidth = 2.0 // 画笔粗细
    ..style = PaintingStyle.stroke; // 用于绘制点时
      PaintingStyle 值无效

  /// 重写绘制内容方法
  @override
  void paint(Canvas canvas, Size size) {
    // 新建一个 path 移动到一个位置，然后画各种线
    Path path = new Path()..moveTo(100.0, 100.0);
    path.lineTo(200.0, 300.0);
    path.lineTo(100.0, 200.0);
    path.lineTo(150.0, 250.0);
    path.lineTo(150.0, 500.0);
    canvas.drawPath(path, _paint);
  }

  /// 是否需要重绘
  @override
  bool shouldRepaint(CustomPainter oldDelegate) {
    return false;
  }
}
```

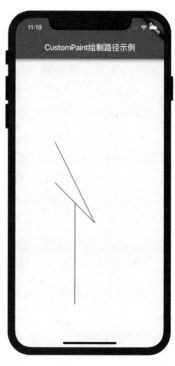

图 11-22　CustomPaint 绘制路径示例

上述示例代码的视图展现大致如图 11-22 所示。

第 12 章 Chapter 12

动 画

随着技术的发展，很多网页开发技术都带有动画效果，比如淡入淡出、渐变、变大变小，等等。Flutter 中的动画效果可以用酷炫来形容，这也是 Flutter 的一大特色。现代的应用程序不仅仅需要程序稳定、好用，还需要好看，体验好。那么动画效果是必不可少的。

本章我们围绕着两个动画组件，带领大家进入 Flutter 动画领域：

❑ AnimatedOpacity（实现渐变效果）。
❑ Hero（页面切换动画）。

12.1 用 AnimatedOpacity 实现渐变效果

在移动端开发中，经常会有一些动画交互，比如淡入淡出。AnimatedOpacity 组件可以自动地在给定的一段时间内改变 child 的透明度。AnimatedOpacity 组件属性如下所示：

属性名	类型	说明
opacity	double	组件透明度。
child	Widget	AnimatedOpacity 子元素。

淡入淡出动画效果如图 12-1 所示。

右图中右下角的 floatingActionButton 按钮会设置 visiable 状态值，opacity 根据 visible 状态值判断会从 0.0 到 1.0 之间取值。左图会慢慢地从隐藏到显示的效果，也就是淡入效果；右图会慢慢从显示到隐藏的效果，也就是淡出效果。完整的示例代码如下所示：

图 12-1　淡入及淡出效果图

```
import 'package:flutter/material.dart';

void main() {
  runApp(new MyApp());
}

class MyApp extends StatelessWidget {

  @override
  Widget build(BuildContext context) {
    final appTitle = "淡入淡出动画示例";
    return new MaterialApp(
      title: appTitle,
      home: new MyHomePage(title:appTitle),
    );
  }
}

class MyHomePage extends StatefulWidget {
  final String title;

  MyHomePage({Key key,this.title}):super(key:key);

  @override
```

```
  _MyHomePageState createState() => new _MyHomePageState();
}

class _MyHomePageState extends State<MyHomePage> {
  //控制动画显示状态变量
  bool _visible = true;

  @override
  Widget build(BuildContext context) {
    return new Scaffold(
      appBar: new AppBar(
        title: new Text(widget.title),
      ),
      body: new Center(
        //添加 Opacity 动画
        child: new AnimatedOpacity(
          //控制 opacity 值 范围从 0.0 到 1.0
          opacity: _visible ? 1.0 : 0.0,
          //设置动画时长
          duration: new Duration(
              milliseconds: 1000
          ),
          child: new Container(
            width: 300.0,
            height: 300.0,
            color: Colors.deepPurple,
          ),
        ),
      ),
      floatingActionButton: new FloatingActionButton(
        onPressed: (){
          //控制动画显示状态
          setState(() {
            _visible = !_visible;
          });
        },
        tooltip: "显示隐藏",
        child: new Icon(Icons.flip),
      ),
    );
  }
}
```

12.2 用 Hero 实现页面切换动画

页面切换时有时需要增加点动画，这样可以增强应用的体验，其中一种作法是可以在页面元素里添加 Hero 组件，就会自带一种过渡的动画效果。

Hero 动画是在两个页面切换过程中发生的。如图 12-2 所示，左侧的页面切换到右侧页面时，有一种推拉的效果。

接下来先写两个页面，各放一张图片。两个页面里的图片都用 Hero 组件包裹，Hero 组件外面再包裹一个 GestureDetector 组件。这样按下图片可以来回切换，切换过程中会有一个过渡效果，如图 12-3 所示。

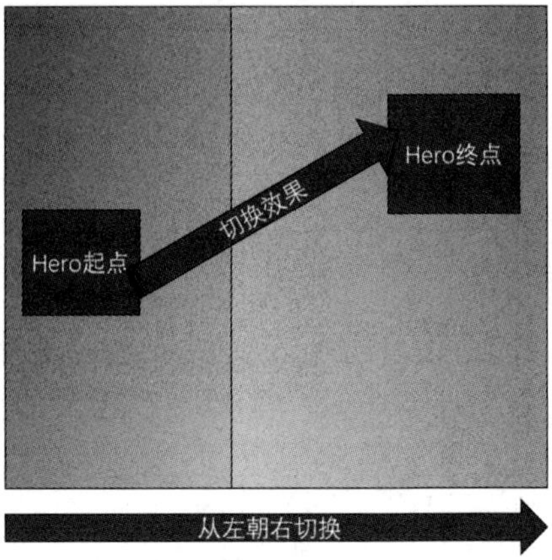

页面切换动画的示例代码如下所示：

```dart
import 'package:flutter/material.dart';

void main() {
  runApp(new MaterialApp(
    title: '页面切换动画',
    home: new FirstPage(),
  ));
}

class FirstPage extends StatelessWidget {

  @override
  Widget build(BuildContext context) {
    return new Scaffold(
      appBar: new AppBar(
        title: new Text("页面切换动画图一"),
      ),
      body: new GestureDetector(
        child: new Hero(
          tag: '第一张图片',
          child: new Image.network(
            "https://timgsa.baidu.com/timg?image&quality=80&size=b9999_10000&sec=1541753399410&di=05760e1c65686b018cf28d440a6ddf5c&imgtype=0&src=http%3A%2F%2Fimg1.cache.netease.com%2Fcatchpic%2FD%2FD7%2FD7D7640C07A00D831EFD2AC270ED7FA7.jpg",
          ),
        ),
        onTap: (){
          Navigator.push(context, new MaterialPageRoute(builder: (_){
            return new SecondPage();
          }));
        },
      ),
    );
  }
}


class SecondPage extends StatelessWidget {
```

图 12-2 Hero 动画

```
@override
Widget build(BuildContext context) {
  return new Scaffold(
    appBar: new AppBar(
      title: new Text("页面切换动画图二"),
    ),
    body: new GestureDetector(

      child: new Hero(
        tag: "第二张图片",
        child: new Image.network("https://timgsa.baidu.com/timg?image&quality=
        80&size=b9999_10000&sec=1541753302014&di=9edfe992f8b9d395134fd977dbfea
        b28&imgtype=0&src=http%3A%2F%2Fimgsrc.baidu.com%2Fimgad%2Fpic%2Fitem%2
        F2f738bd4b31c870143a3a1dc2c7f9e2f0708fff7.jpg"),
      ),

      onTap: (){
        Navigator.pop(context);
      },
    ),
  );
}
```

图 12-3　页面切换动画

第 13 章

Flutter 插件开发

Flutter 插件就是一种 Flutter 的库。只是这个库相对特殊，可以和原生程序打交道。比如调用蓝牙，启用 WIFI，打开手电筒，等等。Flutter 的上层 Dart 语言是无法完成底层操作的，它只能做一些 UI 相关的事情。所以插件开发就显得尤为重要。

插件开发需要涉及两大移动平台的知识，例如：
- iOS：Objectiv-C 及 Swift 语言。
- Android：Java 及 Kotlin 语言。

本章通过案例详细讲解 Flutter 插件是如何开发的。

13.1 新建插件

创建插件有两种方式，一种方式是使用命令行，一种是使用 IDE(AndroidStudio 或 VSCode)。本节我们通过 AndroidStudio 方式来创建一个插件，步骤如下所示。

步骤 1：从菜单里新建一个 Flutter 工程，如图 13-1 所示。

步骤 2：选择 FlutterPlugin 选项，点击 Next，如图 13-2 所示。

步骤 3：输入工程名（即插件

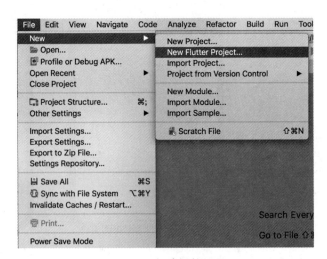

图 13-1　新建插件工程

名称），指定 SDK 路径、工程位置及工程描述。点击 Next，如图 13-3 所示。

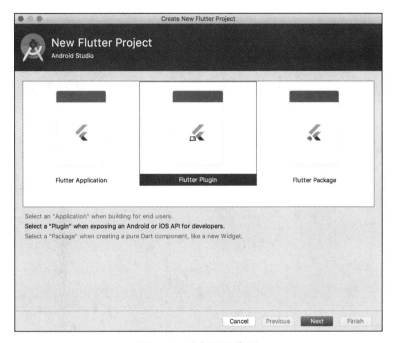

图 13-2　选择工程类型

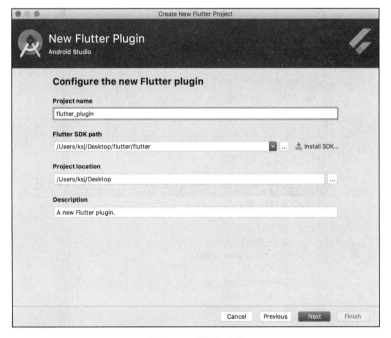

图 13-3　配置工程

步骤4：输入域名通常作为包名使用。在 Platform channel language 一项，根据插件的实际需要进行勾选，默认不用勾选，点击 Finish，如图 13-4 所示。

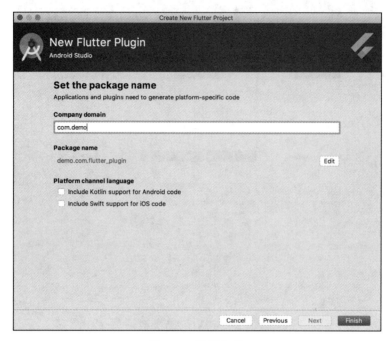

图 13-4　设置包名

步骤5：创建好的插件，和 Flutter 工程一样，也包含 android 和 ios 两个目录，如图 13-5 所示。

图 13-5　插件工程创建完成

13.2 运行插件

1.Android 运行

选择 Android 模拟器,点击运行或者调试按钮即可,如图 13-6 所示。

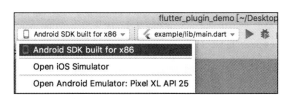

图 13-6　选择 Android 模拟器

模拟器打开后,中间会输出一行文本,表明你运行在 Android 的某个版本之上。这个版本的信息就是读取原生的 API,如图 13-7 所示。

图 13-7　插件在 Android 模拟器上运行

2. iOS 运行

选择 iOS 模拟器,点击运行或者调试按钮即可,如图 13-8 所示。

模拟器打开后,中间会输出一行文本,表明你运行在 iOS 的某个版本之上。这个版本的信息就是读取原生的 API,如图 13-9 所示。

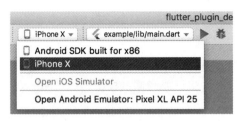

图 13-8　选择 iOS 模拟器

图 13-9　插件在 iOS 模拟器上运行

3. 真机运行

真机和模拟器的运行是一样的，当真机连上开发电脑以后，会在设备列表里出现真机的名称，选择对应的名称，点击 "运行" 或者 "调试" 即可。

有些应用在模拟器上模拟不出来的情况就可以考虑接入真机测试。比如：调用摄像头及麦克风这种情况。

13.3　示例代码分析

本节主要分析如何获取 Android 及 iOS 版本信息，理解插件的通信原理，以及插件上层的 Dart 代码、Android 及 iOS 底层代码是如何编写的。最后再带领大家编写一个插件的测试代码，测试插件功能是否正常。

1. 理解通信原理

Flutter 插件提供 Android 或者 iOS 的底层封装，在 Flutter 层提供组件功能，使 Flutter 可以较方便的调取 Native 的模块。很多平台相关性或者对于 Flutter 实现起来比较复杂的部分，都可以封装成插件。其原理如图 13-10 所示。消息在 client 和 host 之间通过平台通道来进行的，之间的通信都是异步的。

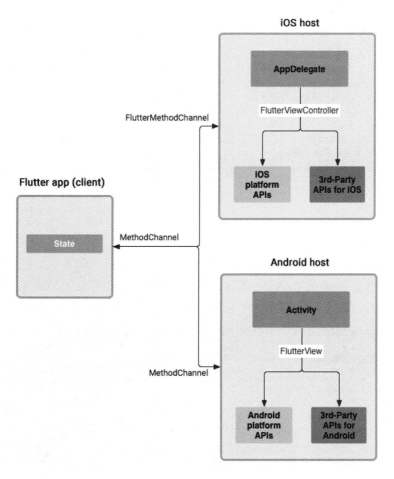

图 13-10　插件通信原理

2. flutter_plugin_demo.dart 插件分析

flutter_plugin_demo.dart 是插件用来对外提供接口的文件。编写步骤如下：

1）首先要导入 services 及 async 库。services 提供与底层通讯的方法，async 作为异步处理。

2）定义通道方法 MethodChannel。注意它的参数为一个标识符，通常为一个字符串，这个字符串必需和原生代码保持一致，否则会导致无法正常通信。

3）编写 platformVersion 方法，此方法用来对外提供服务。注意此方法需要添加异步处理 async/await。使用 channel 的 invokeMethod 方法来调用原生方法获取版本，getPlatformVersion 即为原生方法名。

完整的代码如下所示：

```dart
import 'dart:async';

import 'package:flutter/services.dart';

class FlutterPluginDemo {
  // 定义通道方法 MethodChannel 通道标识这个和原生代码是对应的要保持一致
  static const MethodChannel _channel =
      const MethodChannel('flutter_plugin_demo');

  // 提供外部调用的方法  注意此方法要加异步处理
  static Future<String> get platformVersion async {
    // 调用原生程序获取系统版本
    final String version = await _channel.invokeMethod('getPlatformVersion');
    return version;
  }
}
```

3. Android 部分分析

Android 部分即为 Android 原生模块，代码编写步骤如下所示。

步骤 1：打开插件的 Android 部分源码，不建议直接在主工程里编辑，采用打开 Android 模块的方式。这样在调试时才能断点到原生模块，如图 13-11 所示。

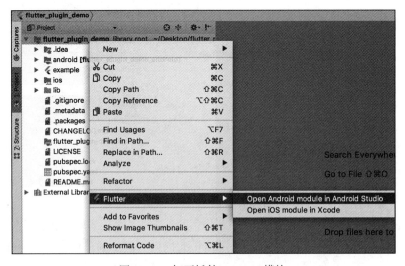

图 13-11　打开插件 Android 模块

步骤 2：打开 FlutterPluginDemoPlugin 文件，路径如图 13-12 所示。

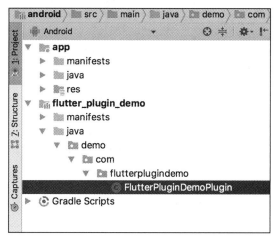

图 13-12　FlutterPluginDemoPlugin 路径

步骤 3：导入通信相关库，如 MethodCall、MethodChannel、Registrar 等，例如导入 MethodCall 如下所示：

```
import io.flutter.plugin.common.MethodCall;
```

步骤 4：新建一个类 FlutterPluginDemoPlugin 继承自 MethodCallHandler，代码如下所示：

```
public class FlutterPluginDemoPlugin implements MethodCallHandler
```

步骤 5：实现插件注册方法，这个方法主要用来定义与上层代码通信的通道。同时设置上层代码调用 handler 方法：

```
MethodChannel(registrar.messenger(), "flutter_plugin_demo");
```

 注意　flutter_plugin_demo 标识符一定要与上层保持一致。

步骤 6：重写 onMethodCall 方法，用来响应上层调用。判断 getPlatformVersion 后获取系统版本并返回，重写方法如下所示：

```
@Override
public void onMethodCall(MethodCall call, Result result)
```

FlutterPluginDemoPlugin 完整实现代码如下所示：

```
package demo.com.flutterplugindemo;

import io.flutter.plugin.common.MethodCall;
```

```
import io.flutter.plugin.common.MethodChannel;
import io.flutter.plugin.common.MethodChannel.MethodCallHandler;
import io.flutter.plugin.common.MethodChannel.Result;
import io.flutter.plugin.common.PluginRegistry.Registrar;

/** FlutterPluginDemoPlugin */
public class FlutterPluginDemoPlugin implements MethodCallHandler {
  /**插件注册 */
  public static void registerWith(Registrar registrar) {
    // 定义与上层代码通信的通道 注意标识符要与上层代码保持一致
    final MethodChannel channel = new MethodChannel(registrar.messenger(),
      "flutter_plugin_demo");
    channel.setMethodCallHandler(new FlutterPluginDemoPlugin());
  }

  @Override
  public void onMethodCall(MethodCall call, Result result) {
    // 判断上层调用的方法
    if (call.method.equals("getPlatformVersion")) {
      // 调用成功后返回系统版本
      result.success("Android " + android.os.Build.VERSION.RELEASE);
    } else {
      result.notImplemented();
    }
  }
}
```

4. iOS 部分分析

iOS 部分即为 iOS 原生模块。代码编写步骤如下所示。

步骤 1：打开插件的 iOS 部分源码，不建议直接在主工程里编辑。采用打开 iOS 模块的方式。这样在调试时才能断点到原生模块。另外在主工程里编写 Objectiv-C 代码也不是很方便，提示不够友好。建议还是使用 Xcode 进行编辑，如图 13-13 所示。

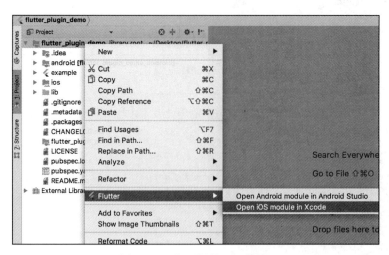

图 13-13　打开插件 iOS 模块

步骤2：打开 FlutterPluginDemoPlugin 文件，路径比较深不太容易寻找。路径如图 13-14 所示。

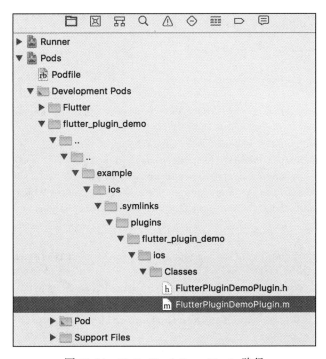

图 13-14　FlutterPluginDemoPlugin 路径

步骤3：导入头文件，代码如下所示：

```
#import "FlutterPluginDemoPlugin.h"
```

步骤4：使用 implementation 实现 FlutterPluginDemoPlugin 接口，代码如下所示：

```
@implementation FlutterPluginDemoPlugin
```

步骤5：实现插件注册方法，这个方法主要是用来定义与上层代码通信的通道，方法体如下所示：

```
+ (void)registerWithRegistrar:(NSObject<FlutterPluginRegistrar>*)registrar {
    //TODO
}
```

> **注意** flutter_plugin_demo 标识符一定要与上层保持一致。

步骤6：编写 handleMethodCall 方法，用来响应上层调用。判断 getPlatformVersion 后获取系统版本并返回，方法体如下所示：

```objc
- (void)handleMethodCall:(FlutterMethodCall*)call result:(FlutterResult)result {
  //TODO
}
```

FlutterPluginDemoPlugin 完整代码如下所示:

```objc
#import "FlutterPluginDemoPlugin.h"

@implementation FlutterPluginDemoPlugin
/** 插件注册 */
+ (void)registerWithRegistrar:(NSObject<FlutterPluginRegistrar>*)registrar {
    // 定义与上层代码通信的通道 注意标识符要与上层代码保持一致
    FlutterMethodChannel* channel = [FlutterMethodChannel
        methodChannelWithName:@"flutter_plugin_demo"
              binaryMessenger:[registrar messenger]];
    FlutterPluginDemoPlugin* instance = [[FlutterPluginDemoPlugin alloc] init];
    [registrar addMethodCallDelegate:instance channel:channel];
}

- (void)handleMethodCall:(FlutterMethodCall*)call result:(FlutterResult)result {
    // 判断上层调用的方法
    if ([@"getPlatformVersion" isEqualToString:call.method]) {
      // 调用成功后返回系统版本
      result([@"iOS " stringByAppendingString:[[UIDevice currentDevice]
        systemVersion]]);
    } else {
      result(FlutterMethodNotImplemented);
    }
}

@end
```

5. 测试插件分析

插件编写完成后,还需要写一个测试工程来验证插件是否有 bug。这个工程命名为 example,如图 13-15 所示。

测试工程编写的步骤如下:

步骤 1:打开 example/pubspec.yaml 文件,添加插件 flutter_plugin_demo 至开发库标签下,如图 13-16 所示。

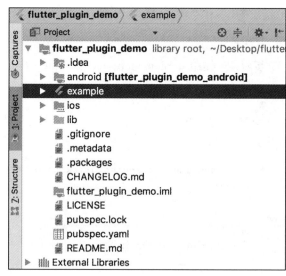

图 13-15 插件测试工程

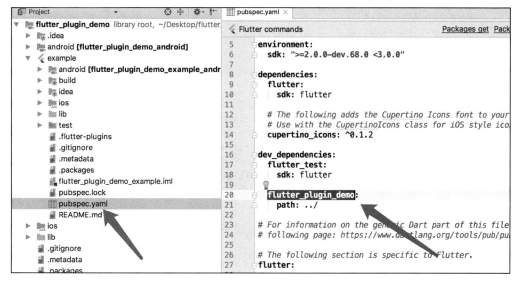

图 13-16　添加插件库至工程配置文件

> **注意** flutter_plugin_demo 是放在 dev_dependencies 标签下的。插件的路径是一个相对路径，如果路径配置错误，会导致找不到插件引用失败错误。

步骤 2：打开 examples/lib/main.dart 文件，编写调用插件方法，调用方法如下所示：

platformVersion = await FlutterPluginDemo.platformVersion;

步骤 3：添加 UI 界面部分代码。运行查看结果。

完整的测试插件代码如下所示：

```
import 'package:flutter/material.dart';
import 'dart:async';

import 'package:flutter/services.dart';
import 'package:flutter_plugin_demo/flutter_plugin_demo.dart';

void main() => runApp(MyApp());

class MyApp extends StatefulWidget {
  @override
  _MyAppState createState() => _MyAppState();
}

class _MyAppState extends State<MyApp> {
  String _platformVersion = 'Unknown';

  @override
  void initState() {
```

```
    super.initState();
    initPlatformState();
}

// 获取底层的方法是异步的,所以调用的方法也要异步处理才行
Future<void> initPlatformState() async {
    String platformVersion;
    // 底层方法有可能会调用失败,所以要使用try/catch语句
    try {
        platformVersion = await FlutterPluginDemo.platformVersion;
    } on PlatformException {
        platformVersion = 'Failed to get platform version.';
    }

    setState(() {
        _platformVersion = platformVersion;
    });
}

@override
Widget build(BuildContext context) {
    return MaterialApp(
        home: Scaffold(
            appBar: AppBar(
                title: const Text('Plugin example app'),
            ),
            body: Center(
                child: Text('Running on: $_platformVersion\n'),
            ),
        ),
    );
}
}
```

上述代码运行的效果请参考上一节"运行插件"。

第 14 章 开发工具及使用技巧

熟练使用工具可以节省大量的开发时间，并提高代码质量。在本章中我们会介绍几款常用的 IDE。从代码的编写，到增加辅助功能并进行调试、性能分析等讲解工具及使用的技巧。

本章涉及的内容有：
- IDE 集成开发环境
- Flutter SDK
- 使用热重载
- 格式化代码
- Flutter 组件检查器

14.1 IDE 集成开发环境

14.1.1 Android Studio / IntelliJ

Flutter 插件在 Android Studio 或 IntelliJ IDE 中提供完全集成的开发体验。所以这两个开发工具的操作基本一致。

首先请参考第 1 章内容安装 Dart 和 Flutter 插件（1.2.1 节"Windows 环境搭建"）。

1. 创建项目

有两种方式创建项目。

第一种，从 Welcome 页面创建，选择 Start a new Flutter project 创建一个新的 Flutter

项目，如图 14-1 所示。

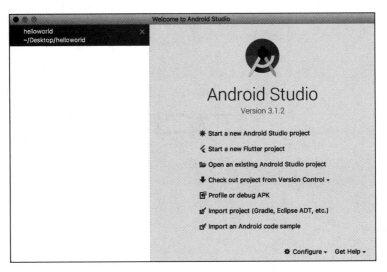

图 14-1　Android Studio 欢迎页面

第二种，从 Android Studio 主界面中依次点击 File → New → New Flutter Project。可以打开创建工程页面，如图 14-2 所示。

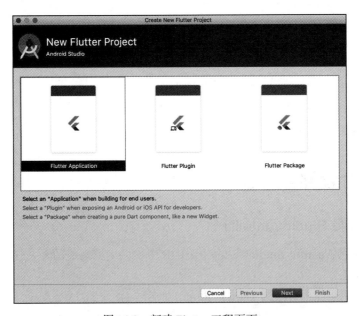

图 14-2　新建 Flutter 工程页面

2. 编辑代码和查看代码问题

Dart 插件可执行代码分析，有以下功能：

❑ 语法高亮显示。
❑ 基于丰富类型分析的代码补全。
❑ 导航到类型声明（Navigate → Declaration），查找类型使用的地方 (Edit → Find → Find Usages)。
❑ 查看当前源代码的所有问题 (View → Tool Windows → Dart Analysis)。任何分析问题将在 Analysis pane 窗口中显示，如图 14-3 所示。

图 14-3　Dart Analysis 图

3. 运行和调试

（1）选择一个 target

运行和调试由主工具栏控制。选择一个 target。在 Android Studio 中打开一个 Flutter 项目时，你应该在工具栏的右侧看到一组 Flutter 特定的按钮，如图 14-4 所示。从左至右依次为目标设备、主程序、运行、调试、覆盖率查看、性能分析、热更新、添加测试到进程、停止。

图 14-4　主工具栏图

> **注意**　如果 Run & Debug 按钮被禁用，并且没有列出任何 target，则说明 Flutter 没有发现任何连接的 iOS 或 Android 设备或模拟器。需要连接设备或启动模拟器才能继续。
> 　　点击 Flutter Target Selector 下拉按钮，这将显示可用的设备列表，选择要运行的设备。当你连接新的设备或启动新的模拟器时，里面会添加新的选项。

（2）无断点运行

点击运行图标，或者调用 Run → Run。底部的 Run 窗格中将会显示日志输出。如图 14-5 所示输出了启动 lib/main.dart 主程序在 iphone X 设备上。

（3）有断点运行

如果需要，可在源代码中设置断点，选中需要设置代码的那一行，在它的左侧点击一下会出现一个小红点，表示设置断点成功。如图 14-6 所示。

点击工具栏的"调试"图标，或者调用 Run → Debug，如图 14-7 所示。

底部的调试窗口将显示调用栈和变量。如图 14-8 所示。在这里可以一步步地朝下测试，查看变量值，查看控制台输出内容。大部分的程序 bug 都是使用这种方式解决的。

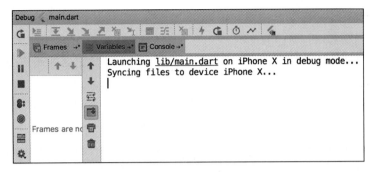

图 14-5　控制台输出

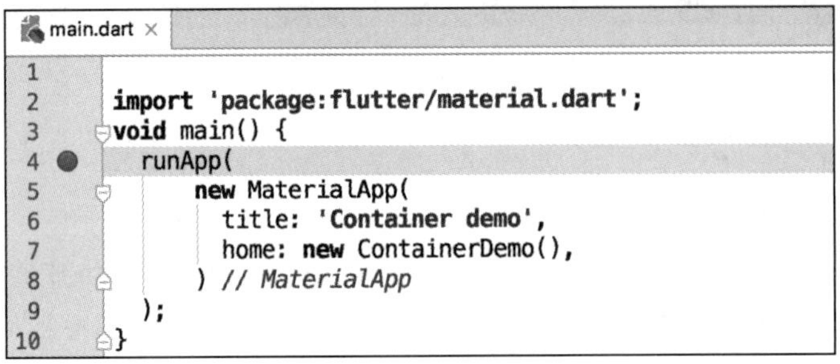

图 14-6　设置断点

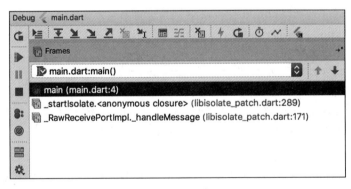

图 14-7　启动 Debugger

图 14-8　调试窗口

4. Observatory 使用

Observatory 是一个附带的基于 HTML 的用户界面的调试和分析工具，在 Debug 窗口点击秒表图标，如图 14-9 所示点击箭头所指图标。IDE 会弹出一个网页，生成一个整个应用程序的详细报告。

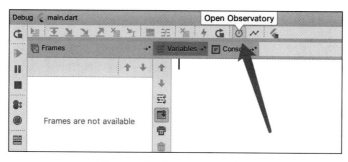

图 14-9 打开 Observatory 窗口

 提示 报告详情请参考：https://dart-lang.github.io/observatory/

5. Flutter 代码提示
（1）辅助 & 快速修正

辅助是与特定代码标识符相关的代码更改。当光标放置在某个组件的标识符上时，会出一个黄色灯泡图标，点击这个灯泡会有一些提示的内容。根据需要选择对应的辅助功能即可，如图 14-10 所示。

图 14-10 辅助功能面板

快速修正是类似的，只有显示一段有错误的代码时，他们可以帮助你纠正它。它用一个红色灯泡表示。

（2）实时模板

实时模板可用于加速输入常用的代码结构块。通常只需要输入一两个字母编辑器就会提示可能要实现的内容。如图14-11所示，ContainerLayout即将添加一个重写build的方法模板。

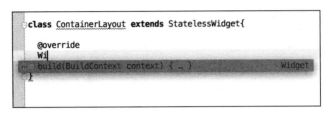

图14-11 实时模板面板

当添加完模板后，近照提示内容实现build方法即可，如图14-12所示。

Flutter插件包含以下模板：

- 前缀stless：创建一个StatelessWidget的子类。
- 前缀stful：创建一个StatefulWidget子类并且关联到一个State子类。
- 前缀stanim：创建一个StatefulWidget子类，并且它关联的State子类包括一个AnimationController。

（3）Android Studio键盘快捷键

在Linux上（Android Studio键盘映射默认为XWin）和Windows键盘快捷键是ctrl-alt-；和ctrl-\。

在macOS上（Android Studio键盘映射Mac OS X 10.5+ copy），键盘快捷键是⌘-?-；和⌘-\。

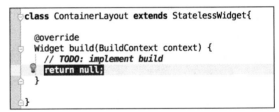

图14-12 模板创建完成

键盘映射可以在设置面板进行更改，选择Keymap，然后在右上角的搜索框中输入flutter。右键单击要更改的绑定并添加键盘快捷键。

快捷键设备如图14-13所示。

（4）热重载 VS 完全重启

热重载通过将更新的源代码文件注入正在运行的Dart VM（虚拟机）中工作。这不仅包括添加新类，还包括向现有类添加方法和字段以及更改现有函数。尽管有几种类型的代码更改无法热重载：

- 全局变量初始化器。
- 静态字段初始化器。
- app的main()方法。

对于这些更改，你可以完全重新启动应用程序，而无需结束调试会话：不要点击停止按钮；只需重新单击运行按钮（如果在运行会话中）或调试按钮（如果在调试会话中），或者按住Shift键并单击"热重载"按钮。

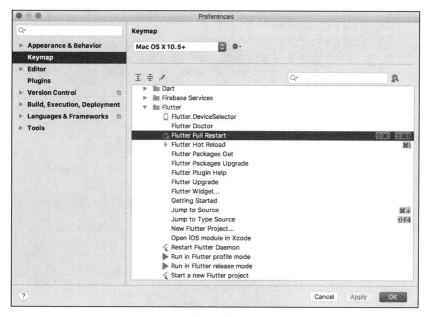

图 14-13　键盘快捷键设置图

6. 在 Android Studio 中编辑 Android 代码

要在 Android Studio 中编辑 Flutter 项目的 Android 代码，可以点击工程右键，找到 Flutter 选择 Open Android module in Android Studio 选项打开，如图 14-14 所示。

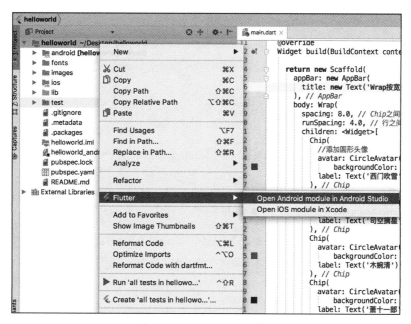

图 14-14　打开 android 模块

此时会弹出一个窗口，选择新建窗口或者当前窗口都可以，如图 14-15 所示。

图 14-15　打开工程窗口

接下来 Android Studio 会下载相关的一些库，请耐心等待，下载完成后会打开 Android 工程的编辑界面，如图 14-16 所示。这样就可以进行 Android 模块的开发了。

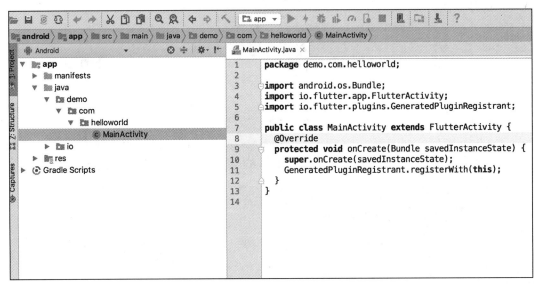

图 14-16　Android 模块编辑界面

> **注意**　打开这个界面有时需要很久的时间，请确保网络通畅。否则由于网络加载的问题导致有些库没有下载下来，会出现很多奇葩的问题。解决起来非常棘手。

7. 在 Xcode 中编辑 iOS 代码

假定你的电脑是 mac。要在 Xcode 中编辑 Flutter 项目的 iOS 代码，可以点击工程右键，找到 Flutter 选择 Open iOS module in Xcode 选项打开，如图 14-17 所示。

Android Studio 会启动 Xcode 并打开 Flutter 下的 iOS 工程，如图 14-18 所示。这样就可以进行 iOS 模块的开发了。

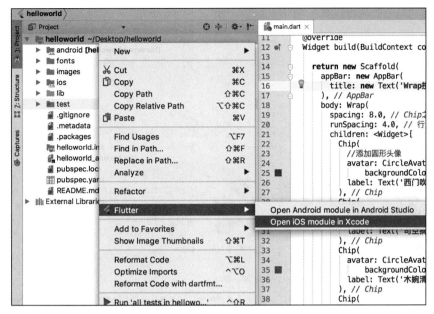

图 14-17　打开 iOS 模块

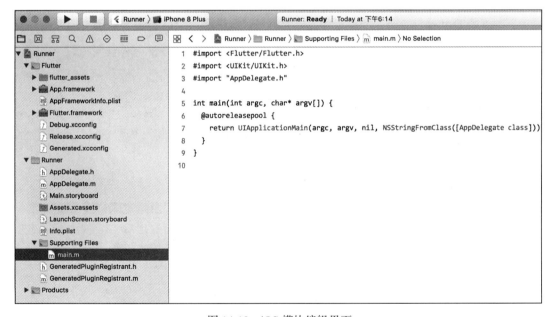

图 14-18　iOS 模块编辑界面

14.1.2　Visual Studio Code

Visual Studio Code（简称 VSCode）是一款免费开源的现代化轻量级代码编辑器。安装好 Flutter 及 Dart 插件后，我们可用来开发 Flutter 应用。

1. 安装和设置

打开 VSCode，点击左侧工具栏的 扩展 按钮，如图 14-19 箭头方向所指。再输入 flutter 后左侧列表面板会出现 flutter 插件，点击安装即可。图中为 禁用 和 卸载 表示已经安装过了，如果没有安装会提示安装。

图 14-19　安装 Flutter 插件

同理安装 Dart 插件也是相同的步骤，无非就是扩展里输入 Dart 即可，如图 14-20 所示。

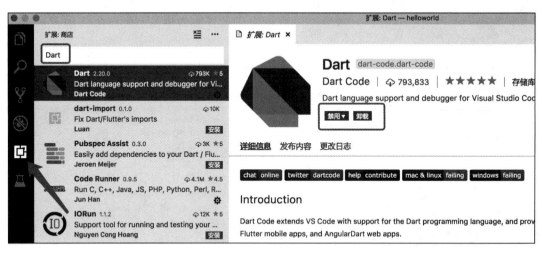

图 14-20　安装 Dart 插件

2. 创建项目

新建一个工程的步骤如下。

步骤 1：打开命令面板 (Ctrl+Shift+P (Cmd+Shift+P on macOS))。如图 14-21 所示。

步骤 2：点击 Flutter:New Project 新建一个工程。如图 14-22 所示。

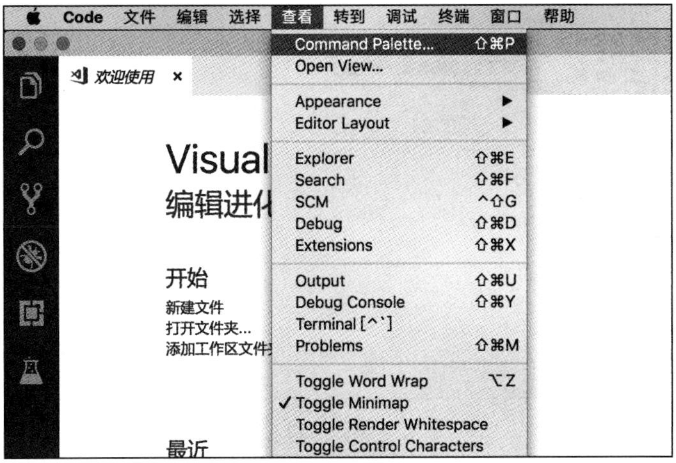

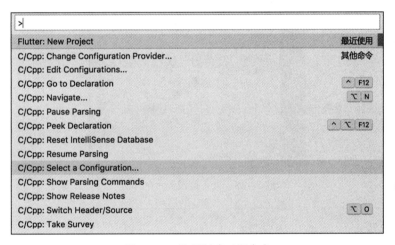

图 14-21　选择打开命令面板菜单

图 14-22　选择新建工程命令

步骤 3：输入工程名，然后回车即可，如图 14-23 所示。

图 14-23　输入工程名

步骤 4：选择工程需要放置的目录，如图 14-24 所示。

步骤 5：等待一会儿，VS Code 会创建一个工程，一开始创建的工程，由于还没有下载好依赖库，所以工程的有些文件夹会有错误颜色显示，如图 14-25 所示，一般为红色。再等待一会儿工程创建完成，就可以编写代码了。

图 14-24　选择工程放置目录

图 14-25　带有错误提示的工程

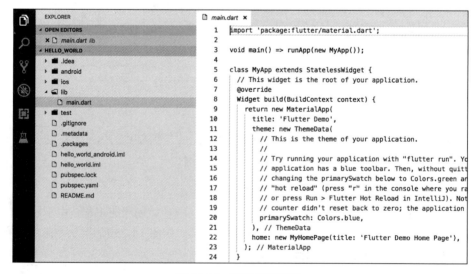

图 14-26　新建好的工程

3. 编辑代码和查看代码问题

Dart 插件执行代码分析，可以：

- 语法高亮显示。
- 基于丰富类型分析的代码补全。
- 导航到类型声明（跳到定义或者 F12），查找类型使用的地方（查找所有的引用或者 Shift+F12）。
- 查看当前源代码的所有问题 (View → Problems 或者 Ctrl+Shift+M (Cmd+Shift+M on macOS))。任何分析问题将在 Analysis pane 窗口中显示，如图 14-27 所示。

图 14-27 问题窗口

4. 运行和调试

点击 Debug → Start Debugging，或者按 F5 可以启动调试程序。

当你从 VS Code 里打开了一个 Flutter 工程，需要观察状态栏的一些特殊信息，包括 Flutter SDK 版本和设备名称（或者找不到设备的信息）。状态栏如图 14-28 所示。VS Code 会自动选择最新的一个设备进行连接，如果你有多个设备/模拟器连接着，点击状态栏设备按钮选择一个设备即可。

图 14-28 VS Code 状态栏信息

> **注意** 如果你没有发现 Flutter 版本或者设备信息，那么你的工程可能还没有被 VS Code 检测成 Flutter 工程。请确保在你的工程文件夹下要有 pubspec.yaml 文件。如果状态栏里的信息显示没有设备，没有连接 iOS 或者 Android 设备或者模拟器。那么你需要连接一台设备或者启动一个模拟器，然后重启。

（1）无断点运行

在 IDE 的窗口里点击 Debug → Start，或者按下 Ctrl+F5，状态栏会显示一个橙色的信息表示你已经启动调试程序，如图 14-29 所示。

（2）有断点运行

在源代码中设置断点，如图 14-30 所示代码左侧几个小点，就表示设置好了断点。

图 14-29 调试控制台面板

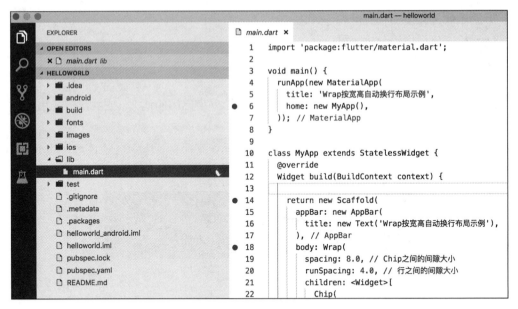

图 14-30　设置断点

在 IDE 的窗口里点击 Debug → Start Debugging，或者按下 F5。可以启动调试面板，如图 14-31 所示。其中左侧的 Debug Sidebar 显示栈及变量信息，底部的 Debug Console 显示日志信息。

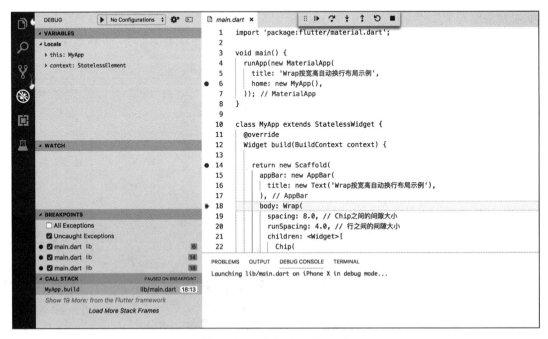

图 14-31　VS Code 调试面板

5. Flutter 代码提示

（1）辅助 & 快速修正

辅助功能是与特定代码标识符相关的代码更改。当光标放置在某个组件的标识符上时，会出一个黄色灯泡图标，点击这个灯泡会有一些提示的内容。根据需要选择对应的辅助功能即可，如图 14-32 所示。

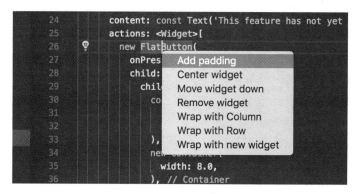

图 14-32　代码提示

快速修正功能类似，当一段代码有错误时，可以帮助你纠正它。错误代码用红色波浪线显示。

（2）实时模板

实时模板有助于加速输入常用的代码结构块。仅输入"前缀"就可调用代码块，然后在代码补全窗口中选择即可。如图 14-33 所示，当输入 bu 马上提示是否插入一个 build 方法的模板。

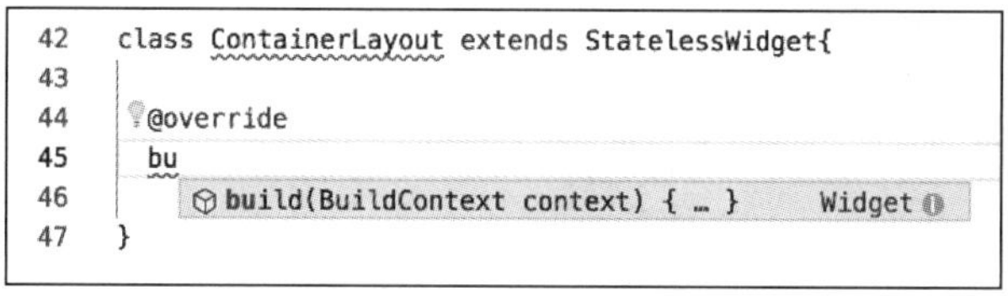

图 14-33　实时模板面板

当代码补充完以后如图 14-34 所示，只需要继续实现 build 方法即可。

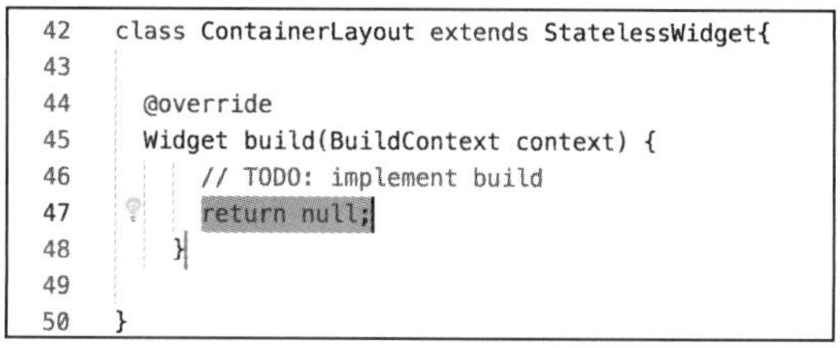

图 14-34　实时模板补充完成

Flutter 插件包含以下前缀：
- stless：创建一个 StatelessWidget 子类。
- stful：创建一个 StatefulWidget 子类并且关联到一个 State 子类。
- stanim：创建一个 StatefulWidget 子类，且它关联的 State 子类包括 AnimationController。

（3）热重载 VS 完全重启

热重载通过将更新的源代码文件注入正在运行的 Dart VM（虚拟机）中工作。这不仅包括添加新类，还包括向现有类添加方法和字段以及更改现有函数。但是有几种类型的代码更改无法热重载：
- 全局变量初始化器。
- 静态字段初始化器。
- app 的 main() 方法。

对于这些更改，你可以完全重新启动应用程序，而无需结束调试会话：不要点击停止按钮；只需重新单击运行按钮（如果在运行会话中）或调试按钮（如果在调试会话中），或者按下 Ctrl+F5 键。

14.2　Flutter SDK

Flutter SDK 更新非常频繁，在开发过程中强烈建议你使用 Releases 版本。下载地址为：https://github.com/flutter/flutter/releases。选择最新版本即可。

1. 项目指定 Flutter SDK

你可以在 pubspec.yaml 文件中指定 Flutter SDK 的环境。例如，以下代码片段指定 Flutter SDK 在一个版本范围内：

```
name: helloworld
description: A new Flutter application.

# The following defines the version and build number for your application.
# A version number is three numbers separated by dots, like 1.2.43
# followed by an optional build number separated by a +.
# Both the version and the builder number may be overridden in flutter
# build by specifying --build-name and --build-number, respectively.
# Read more about versioning at semver.org.
version: 1.0.0+1

environment:
  sdk: ">=2.0.0-dev.68.0 <3.0.0"

dependencies:
  flutter:
```

2. 升级 Flutter channel 和 Packages

要同时更新 Flutter SDK 及其依赖包，在你的应用程序根目录（包含 pubspec.yaml 文件的目录）中运行 flutter upgrade 命令：

```
$ flutter upgrade
```

14.3　使用热重载

Flutter 的热重载功能可以帮助你在无需重新启动应用的情况下快速、轻松地进行测试、构建用户界面、添加功能以及修复错误。当你更新了代码后并且进行保存，IDE 就自动进行重载。Android Studio 的热重载的效果如图 14-35 所示。

图 14-35　Android Studio 热重载

VSCode 的热重载的效果如图 14-36 所示，VSCode 可以点击刷新按钮。

如果你正在使用命令行 flutter run 运行应用程序，热重载后，你将在控制台中看到类似于如下内容的消息：

```
Performing hot reload...
Reloaded 1 of 448 libraries in 2,777ms.
```

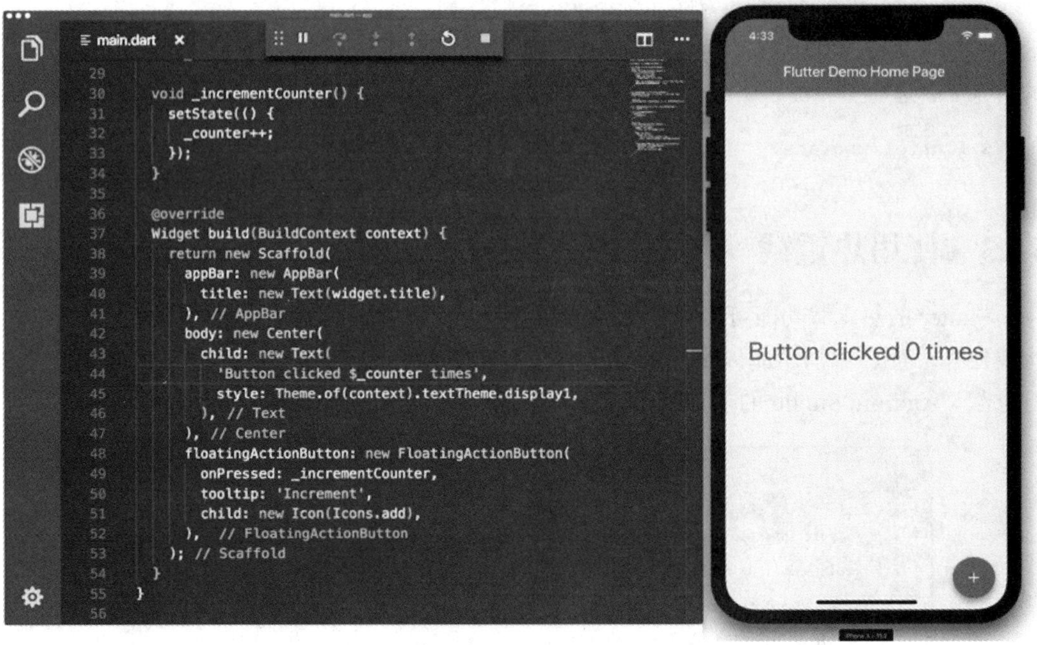

图 14-36　VSCode 热重载

当代码更改后引入了编译错误时，热重载会生成类似于如下内容的错误消息。在这种情况下，只需纠正代码错误，就可以继续使用热重载：

```
Hot reload was rejected:
'/Users/obiwan/Library/Developer/CoreSimulator/Devices/AC94F0FF-16F7-46C8-
B4BF-218B73C547AC/data/Containers/Data/Application/4F72B076-42AD-44A4-A7CF-
57D9F93E895E/tmp/ios_testWIDYdS/ios_test/lib/main.dart': warning: line 16 pos
38: unbalanced '{' opens here
  Widget build(BuildContext context) {
                                     ^
'/Users/obiwan/Library/Developer/CoreSimulator/Devices/AC94F0FF-16F7-46C8-
B4BF-218B73C547AC/data/Containers/Data/Application/4F72B076-42AD-44A4-A7CF-
57D9F93E895E/tmp/ios_testWIDYdS/ios_test/lib/main.dart': error: line 33 pos 5:
unbalanced ')'
    );
    ^
```

14.4　格式化代码

格式化代码分两种：在 Android Studio 和 IntelliJ 中进行，在 VSCode 中进行。在 Android Studio 和 IntelliJ 中格式化代码的步骤如下：

1）选中要格式化的代码，如图 14-37 所示。

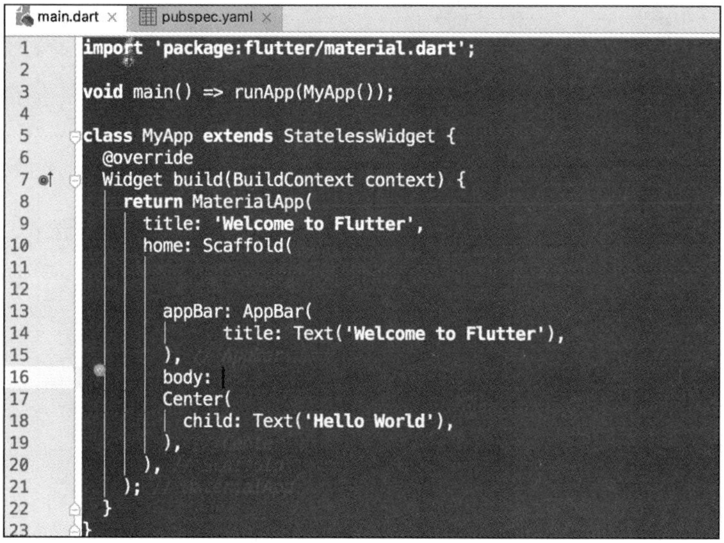

图 14-37 选中要格式化的代码

2）点击 Code → Reformat Code 格式化代码，如图 14-38 所示。

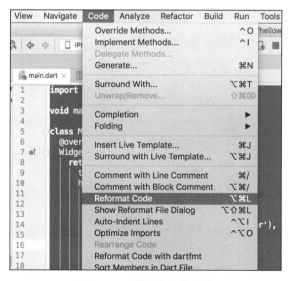

图 14-38 代码格式化菜单

3）格式化完成如图 14-39 所示。请仔细观察步骤 1 及步骤 3 中的代码，看看有什么变化没有。

在 VSCode 中格式代码时，首先选中代码，然后请右键单击代码窗口并选择"格式化文件"，如图 14-40 所示。要在保存文件时自动格式化代码，请将 editor.formatOnSave 设置为 true。

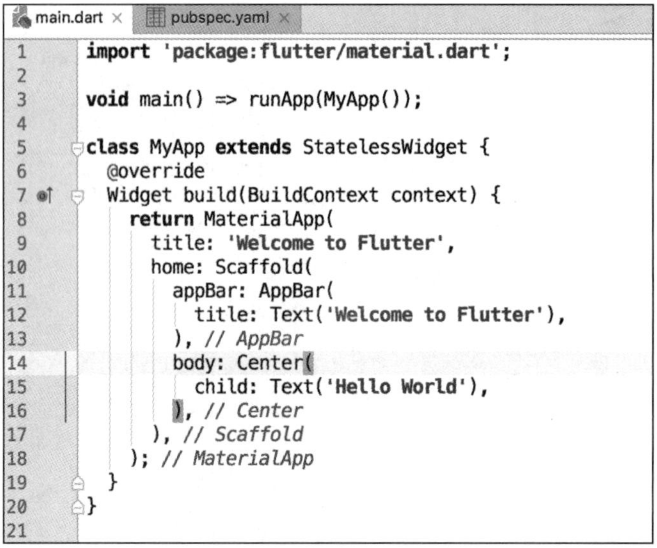
main.dart
pubspec.yaml
import 'package:flutter/material.dart';
void main() => runApp(MyApp());
class MyApp extends StatelessWidget {
@override
Widget build(BuildContext context) {
return MaterialApp(
title: 'Welcome to Flutter',
home: Scaffold(
appBar: AppBar(
title: Text('Welcome to Flutter'),
), // AppBar
body: Center(
child: Text('Hello World'),
), // Center
), // Scaffold
); // MaterialApp
}
}

图 14-39　代码格式化完成

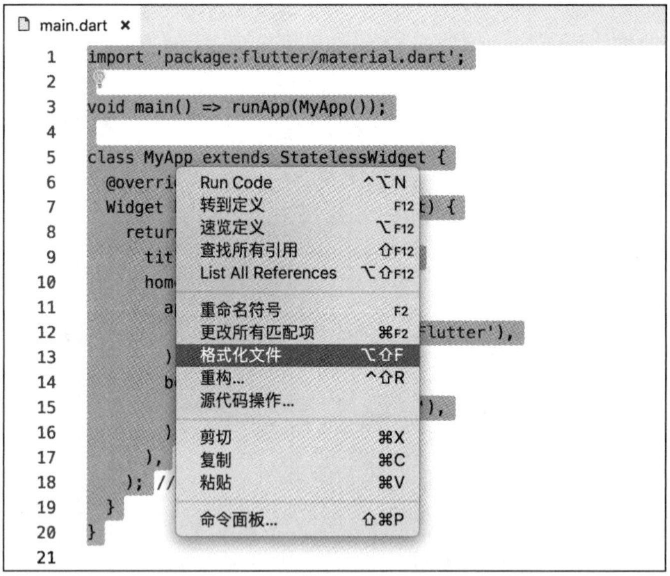
main.dart
import 'package:flutter/material.dart';
void main() => runApp(MyApp());
class MyApp extends StatelessWidget {
Run Code ^⌥N
转到定义 F12
速览定义 ⌥F12
查找所有引用 ⇧F12
List All References ⌥⇧F12
重命名符号 F2
更改所有匹配项 ⌘F2
格式化文件 ⌥⇧F
重构... ^⇧R
源代码操作...
剪切 ⌘X
复制 ⌘C
粘贴 ⌘V
命令面板... ⇧⌘P

图 14-40　VSCode 格式化代码

14.5　Flutter 组件检查器

Flutter Widget Inspector 是用于浏览 Flutter Widget 的强大工具。这款工具类似于浏览器里的开发者工具。Inspector 目前可用于 Android Studio 或 IntelliJ IDEA 的 Flutter 插件。

Flutter 框架使用 Widget 作为核心构建块，从组件（文本、按钮、toggle 等）到布局（居中、填充、行、列等）的任何内容都是。Inspector 是用于可视化和浏览 Flutter 组件的强大工具。如图 14-41 所示。在以下情况下使用这个工具可能会有帮助：

❑ 不清楚现有布局。
❑ 诊断布局问题。

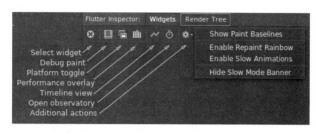

图 14-41　Flutter Widget Inspector 窗口

点击 Flutter Inspector 工具栏上的 Select widget，然后点击设备（真机或虚拟机）以选择一个 Widget。所选 Widget 将在设备和 Widget 树中高亮显示。如图 14-42 所示，当点击左侧的 Text 节点时，右侧的模拟器中文本 "0" 处于选中状态。同样，当鼠标移动到右侧的组件上时，左侧的检查工具也会处于选中状态。

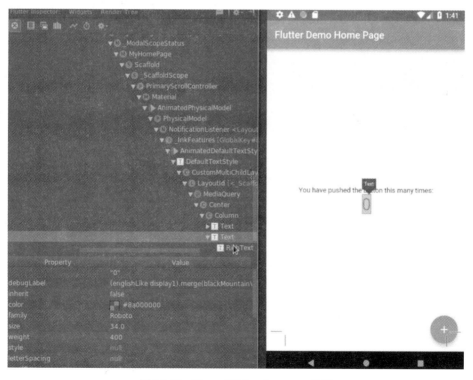

图 14-42　Flutter Widget Inspector 操作

然后，你可以浏览 IDE 中的交互式 Widget 树，以查看附近的 Widget 并查看其字段值。如果你想调试布局问题，那么 Widgets 树可能不够详细。在这种情况下，单击 Render Tree 选项卡查看树中相同位置的渲染树。如图 14-43 所示。当调试布局问题时，关键是看 size 和 constraints 字段。约束沿着树向下传递，尺寸向上传递。

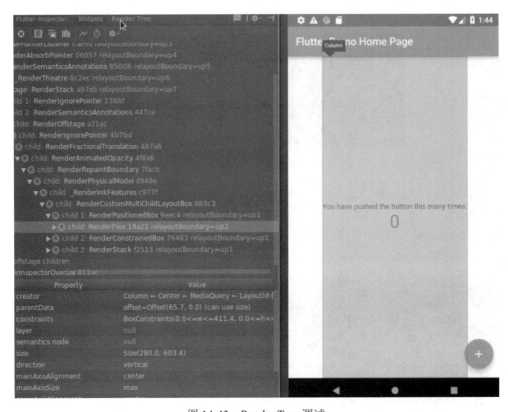

图 14-43　Render Tree 调试

第 15 章

测试与发布应用

前面的章节主要讲解了开发相关的知识。程序开发完成后，首先就是要进行测试，然后构建打包发布应用。所以本章主要讲述的是软件应用开发的后续环节。

本章围绕以下几个方面给大家讲解：
❏ 测试应用
❏ 发布 Android 版 App
❏ 发布 iOS 版 App

15.1 测试应用

15.1.1 简介

如果你的应用较小，手动测试即可。比如应用只有几个页面的情况。当功能页面达到几十个时，手动测试的难度就变大了。一套完整的自动化测试将有助于确保应用在发布之前正确执行，同时快速修复错误。

有很多种自动化测试方法。总结如下：
❏ **单元测试**：测试单一功能、方法或类。例如，将测单元的外部依赖性模拟出来，如 package:mockito。单元测试通常不会读取/写入磁盘、渲染到屏幕，也不会从运行测试的进程外部接收用户操作。单元测试的目标是在各种条件下验证逻辑单元的正确性。
❏ **组件测试**：测试单个 Widget，目标是验证 Widget 如预期的外观和交互功能。测试 Widget 涉及多个类，并且需要提供适当的 Widget 生命周期上下文的测试环境。例如，它应该能够接收和响应用户操作和事件，执行布局并实例化子 Widget。Widget 测试比单元测试更全面。
- **集成测试**：测试整个应用程序或应用程序的很大一部分。通常，集成测试可以在真实设备或模拟器上运行，例如 iOS Simulator 或 Android Emulator。集成测试的目标是验证应用程序作为一个整体是否正确运行，它所组成的所有 Widget 如预期的那样相互集成。还可以使用集成测试来验证应用的性能，比如在使用到音视频的场景下就需要详细测试性能。

表 15-1 总结了在不同类型测试之间进行选择的权衡因素。

表 15-1 不同测试类型对比

类目	单元测试	Widget 测试	集成测试
维护成本	低	高	很高
依赖	少	多	很多
执行速度	快	慢	非常慢

> 提示　经过充分测试的应用程序，往往经过非常多的单元测试和 Widget 测试，通过代码覆盖进行跟踪，以及覆盖所有重要使用场景的大量集成测试。

15.1.2　单元测试

某些 Flutter 库，如 dart:ui，在独立的 Dart VM 附带的 Dart SDK 的中是不可用。flutter test 命令允许你在本地 Dart VM 中运行测试，使用无头版（不会显示 UI）的 Flutter 引擎。使用这个命令你可以运行任何测试，不管它是否依赖 Flutter 的库。

使用 package:test，编写一个 Flutter 单元测试。package:test 文档地址在这里：https://github.com/dart-lang/test/blob/master/README.md。

例如，将此文件添加到 test/unit_test.dart，代码如下所示：

```
import 'package:test/test.dart';

void main() {
  test('my first unit test', () {
    var answer = 42;
    expect(answer, 42);
  });
}
```

另外，必须将以下内容添加到 pubspec.yaml：

```
dev_dependencies:
  flutter_test:
    sdk: flutter
```

即使测试本身没有明确导入 flutter_test，也需要这样做，因为测试框架本身在后台也使用了它。

要运行测试,从你的项目目录(而不是从 test 子目录)运行 flutter test test/unit_test.dart。要运行所有测试,请从项目目录运行 flutter test。

15.1.3 Widget 测试

类似于单元测试,实现 Widget 测试要在测试中执行与 Widget 的交互,请使用 Flutter 提供的 WidgetTester。例如,可以发送点击和滚动手势,还可以使用 WidgetTester 在 Widget 树中查找子 Widget、读取文本、验证 Widget 属性的值是否正确。

例如,新建一个 widget_test.dart 文件,将此文件添加到 test/widget_test.dart。代码如下所示:

```
import 'package:flutter/material.dart';
import 'package:flutter_test/flutter_test.dart';

void main() {
  testWidgets('my first widget test', (WidgetTester tester) async {
    // 你可以使用 Keys 来定位需要测试的组件
    var sliderKey = new UniqueKey();
    var value = 0.0;

    // 传递一个 UI 组件给测试对象
    await tester.pumpWidget(
      new StatefulBuilder(
        builder: (BuildContext context, StateSetter setState) {
          return new MaterialApp(
            home: new Material(
              child: new Center(
                child: new Slider(
                  key: sliderKey,
                  value: value,
                  onChanged: (double newValue) {
                    setState(() {
                      value = newValue;
                    });
                  },
                ),
              ),
            ),
          );
        },
      ),
    );
    expect(value, equals(0.0));

    // Taps on the widget found by key
    await tester.tap(find.byKey(sliderKey));
```

```
        // 验证组件更新的值是否正确
        expect(value, equals(0.5));
    });
}
```

运行 flutter test test/widget_test.dart。查看所有可用于 widget 测试的 package:flutter_test API。

为了帮助调试 Widget 测试，你可以使用 debugDumpApp() 函数对测试的 UI 状态进行可视化，或者只是简单地在首选运行时环境（例如模拟器或设备）中运行 flutter run test/widget_test.dart 以查看测试运行。在运行 flutter run 进行测试的会话期间，还可以点击 Flutter 工具的部分屏幕来打印建议的 Finder。

15.1.4 集成测试

也许你熟悉 Selenium/WebDriver(Web)、Espresso(Android) 或 UI Automation(iOS) 集成测试工具，Flutter Driver 也是集成测试工具。此外，Flutter Driver 还提供 API 以跟踪测试执行的操作的性能。

Flutter Driver 包括：
- 一个命令行工具 flutter drive
- 一个包 package:flutter_driver (API)

Flutter Driver 的功能包括：
- 创建指令化的应用程序
- 写一个测试
- 运行测试

下面介绍用 Flutter Driver 进行集成测试的步骤。

1. 添加 flutter_driver 依赖项

要使用 flutter_driver，必须将以下代码块添加到你的 pubspec.yaml：

```
dev_dependencies:
  flutter_driver:
    sdk: flutter
```

2. 创建指令化的 Flutter 应用程序

一个指令化的应用程序是一个 Flutter 应用程序，它启用了 Flutter Driver 扩展。启用扩展请调用 enableFlutterDriverExtension()。

例如，你有一个入口点的应用程序 my_app/lib/main.dart。要创建它的指令化版本，请在 my_app/test_driver/ 下创建一个 Dart 文件。在你正在测试的功能之后命名它；接下来定位到 my_app/test_driver/user_list_scrolling.dart。代码如下所示：

```dart
// 这一行导入扩展
import 'package:flutter_driver/driver_extension.dart';

void main() {
  // 启用扩展
  enableFlutterDriverExtension();

  // 调用 main() 程序或调用 runApp 方法，测试你想要测试的程序
}
```

3. 编写集成测试

集成测试是一个简单的 package:test 测试，它使用 Flutter Driver API 告诉应用程序执行什么操作，然后验证应用程序是否执行了此操作。

例如，让我们的测试记录下性能跟踪（performance timeline）。创建一个 user_list_scrolling_test.dart 测试文件位于 my_app/test_driver/ 下。代码如下所示：

```dart
import 'dart:async';

// 导入 Flutter Driver API
import 'package:flutter_driver/flutter_driver.dart';
import 'package:test/test.dart';

void main() {
  group('scrolling performance test', () {
    FlutterDriver driver;

    setUpAll(() async {
      // 连接 app
      driver = await FlutterDriver.connect();
    });

    tearDownAll(() async {
      if (driver != null) {
        // 关闭连接
        driver.close();
      }
    });

    test('measure', () async {
      // 记录闭包中的 performance timeline
      Timeline timeline = await driver.traceAction(() async {
        // 找到滚动的 user list
        SerializableFinder userList = find.byValueKey('user-list');

        // 向下滚动 5 次
        for (int i = 0; i < 5; i++) {
```

```
    // 向下滚动 300 像素，使用 300 毫秒
    await driver.scroll(
        userList, 0.0, -300.0, new Duration(milliseconds: 300));

    // 模拟用户的手势，延长 500 毫秒进行下一次滚动
    await new Future<Null>.delayed(new Duration(milliseconds: 500));
    }

    // 向上滚动 5 次
    for (int i = 0; i < 5; i++) {
      await driver.scroll(
          userList, 0.0, 300.0, new Duration(milliseconds: 300));
      await new Future<Null>.delayed(new Duration(milliseconds: 500));
    }
    });

    // timeline 对象包含了滚动中的性能方面数据，比如平均每侦的渲染时间，很有用
    TimelineSummary summary = new TimelineSummary.summarize(timeline);
    summary.writeSummaryToFile('stocks_scroll_perf', pretty: true);
    summary.writeTimelineToFile('stocks_scroll_perf', pretty: true);
    });
  });
}
```

4. 运行集成测试

要在 Android 设备上运行测试，请通过 USB 将设备连接到计算机并启用 USB 调试。然后运行以下命令：

```
flutter drive --target=my_app/test_driver/user_list_scrolling.dart
```

该命令将：

- 构建 --target 应用，并将其安装在设备上。
- 启动应用。
- 运行 my_app/test_driver/ 下的 user_list_scrolling_test.dart。

你可能想知道该命令如何找到正确的测试文件。flutter drive 命令使用一种约定来查找测试文件，这种测试文件与 --target 应用程序在同一目录中具有相同文件名但具有 _test 后缀。

15.2 发布 Android 版 App

在开发过程中，当你使用 flutter run 命令行或者 Android Studio 中通过工具栏运行和调试按钮进行测试。默认情况下，Flutter 构建的是 debug 版本。

当你要为 Android 准备发布版时，例如要发布到应用商店，请按照如下步骤操作。

接下来的示例中，以我的 helloworld 工程 (在第 1 章第 1.3 节) 为例。步骤中提到的 helloworld 工程最终换成你的实际工程名即可。

15.2.1 检查 App Manifest

查看默认应用程序清单文件 (路径请参考 helloworld 工程 AndroidManifest.xml 文件路径，如图 15-1 所示)，并验证这些值是否正确，特别需要重点检查如下几项：

- application：编辑 application 标签，这是应用的名称。
- uses-permission：是否开启相关权限，比如网络。
- 其他硬件相关的权限，比如摄像头、麦克风、蓝牙等。

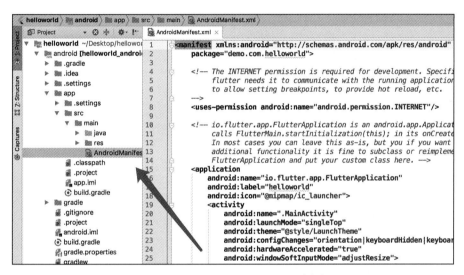

图 15-1　AndroidManifest.xml 路径

 提示　应用程序清单请访问如下地址：https://developer.android.com/guide/topics/manifest/manifest-intro

15.2.2 查看构建配置

Android 开发采用的是 Gradle 构建文件 build.gradle，文件路径如图 15-2 所示。验证这些值是否正确，特别需要重点检查如下几项：

- applicationId：指定始终唯一的 appid，如：demo.com.helloworld。
- versionCode：指定应用程序版本号。
- versionName：指定应用程序版本号字符串。
- minSdkVersion：指定最低的 API 级别。
- targetSdkVersion：指定应用程序设计运行的 API 级别。

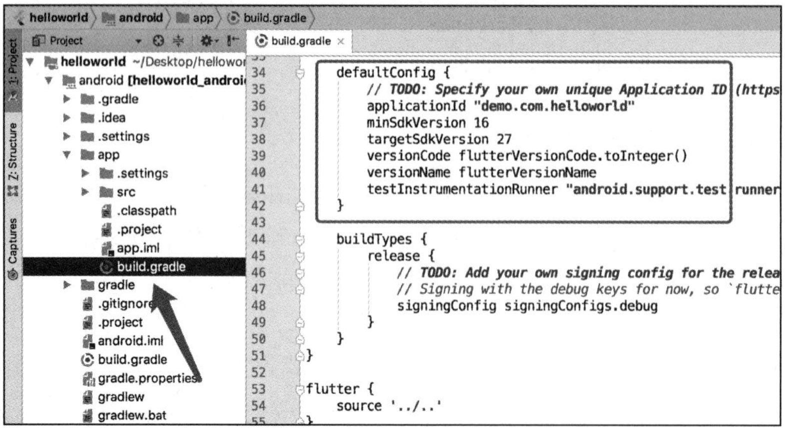

图 15-2　构建配置文件 build.gradle 路径

15.2.3　添加启动图标

所有的 App 应用安装在真机上都要有一个图标，图标通常为产品的 logo。从以下几个方面自定义此图标：

1）查看 Android 启动图标设计指南，然后创建图标。文档地址为：https://material.io/design/iconography/。

2）在 helloworld/android/app/src/main/res/ 目录中（如图 15-3 所示），每个目录都添加一个启动图标并统一命名。建议让美工分别做几种分辨率的图标，以适应不同 DPI 手机的展示需要。

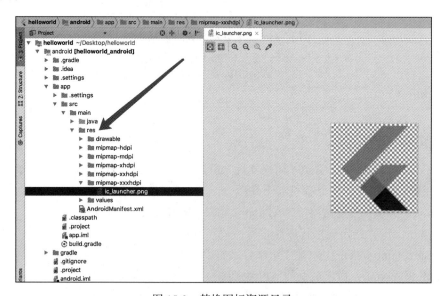

图 15-3　替换图标资源目录

3）在 AndroidManifest.xml 中，将 application 标记的 android:icon 属性更新为引用上一步中的图标（例如 <application android:icon="@mipmap/ic_launcher" ...）。如图 15-4 所示。

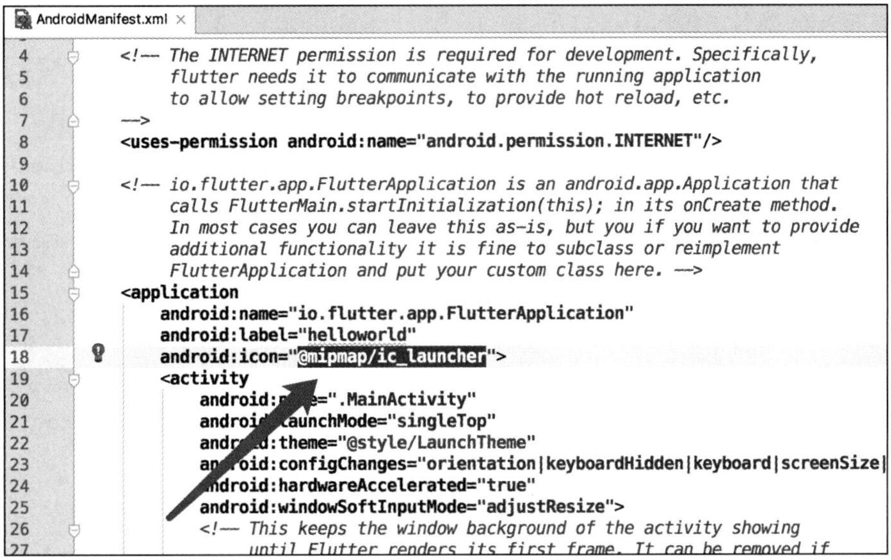

图 15-4　更新属性

4）要验证图标是否已被替换，请运行你的应用程序并检查应用图标。图标最终的安装效果如图 15-5 所示。请核对应用名及图标样式。

图 15-5　安装后的图标

15.2.4 App 签名

App 签名的目的是确保 App 的安装包来自于原创的作者,且 App 没有被篡改。所以这里需要添加 App 签名。具体步骤如下:

步骤 1:创建 keystore。如果你有现成的 keystore,请跳至下一步。如果没有,请通过运行以下命令来创建一个:

```
keytool -genkey -v -keystore ~/key.jks -keyalg RSA -keysize 2048 -validity 10000 -alias key
```

> **注意** 保持文件私密;不要将它加入到公共源代码控制中。keytool 可能不在你的系统路径中。它是 Java JDK 的一部分,它是作为 Android Studio 的一部分安装的。有关具体路径,请查找 JavaJDK 相关知识。

步骤 2:引用应用程序中的 keystore。创建一个名为 helloworld/android/key.properties 的文件,其中包含对密钥库的引用:

```
storePassword=<password from previous step>
keyPassword=<password from previous step>
keyAlias=key
storeFile=<location of the key store file, e.g. /Users/<user name>/key.jks>
```

> **注意** 保持文件私密,不要将它加入公共源代码控制中。

步骤 3:在 Gradle 中配置签名。通过编辑 helloworld/android/app/build.gradle 文件为你的应用配置签名。

将如下内容:

```
android {
```

替换为:

```
def keystorePropertiesFile = rootProject.file("key.properties")
def keystoreProperties = new Properties()
keystoreProperties.load(new FileInputStream(keystorePropertiesFile))

android {
```

将如下内容:

```
buildTypes {
  release {
    // TODO: 添加你的 release 版的鉴名配置,现在可以使用 debug 版的 key
    // flutter run --release 命令可以使用了
    signingConfig signingConfigs.debug
  }
}
```

替换为：

```
signingConfigs {
  release {
    keyAlias keystoreProperties['keyAlias']
    keyPassword keystoreProperties['keyPassword']
    storeFile file(keystoreProperties['storeFile'])
    storePassword keystoreProperties['storePassword']
  }
}
buildTypes {
  release {
    signingConfig signingConfigs.release
  }
}
```

现在，你的应用的发布版本将自动进行签名。

15.2.5 构建发布版 APK 并安装在设备上

有了签名后，你就可以构建发布版（release）APK。以工程名为 helloworld 为例，使用命令行：

❑ cd helloworld (helloworld 为你的工程目录)。

❑ 运行 flutter build apk (flutter build 默认会包含 --release 选项)。

打包好的发布 APK 位于 helloworld/build/app/outputs/apk/app-release.apk。

然后，按照以下步骤在已连接的 Android 设备上安装 APK：

使用命令行：

❑ 用 USB 将 Android 设备连接到你的电脑。

❑ 录入 cd helloworld。

❑ 运行 flutter install。

最后将应用的发布版发布到 Google Play 商店。发布的详细说明请参考如下链接的内容：https://developer.android.com/distribute/best-practices/launch/。

15.3 发布 iOS 版 App

本节会一步步帮你将 Flutter 应用程序发布到 App Store。接下来以 helloworld 工程为例，步骤中提到的 helloworld 工程最终换成你的实际工程名即可。

15.3.1 准备工作

苹果的应用审核非常严格，所以在开始发布你的应用程序之前，请确保它符合 Apple 的 App Review Guidelines。为了将你的应用发布到 App Store，你需要注册一个苹果开发者帐号。

提示 App Review Guidelines 请访问 https://developer.apple.com/app-store/review/。苹果开发者中心请访问 https://developer.apple.com/。

15.3.2 在 iTunes Connect 上注册应用程序

iTunes Connect 是你管理应用程序生命周期的地方，需要在 App Store 定义你的应用程序名称和说明，添加屏幕截图，设置价格并管理版本。

注册你的应用程序涉及两个步骤：注册唯一的 Bundle ID，并在 iTunes Connect 上创建应用程序记录。

有关 iTunes Connect 的详细概述，请参阅 iTunes Connect 开发者指南。

  提示 iTunes Connect：https://developer.apple.com/support/app-store-connect/
iTunes Connect 开发者指南：https://developer.apple.com/support/app-store-connect/

15.3.3 注册一个 Bundle ID

每个 iOS 应用程序都与一个 Bundle ID 关联，这是一个在 Apple 注册的唯一标识符。要为你的应用注册一个 Bundle ID，请按照以下步骤操作。

步骤 1：在 Xcode 系统配置的 Accounts 选项卡中，可以打开开发者帐户的 Apple IDs 页。点击左下角 + 号创建一个 Bundle ID，如图 15-6 所示。

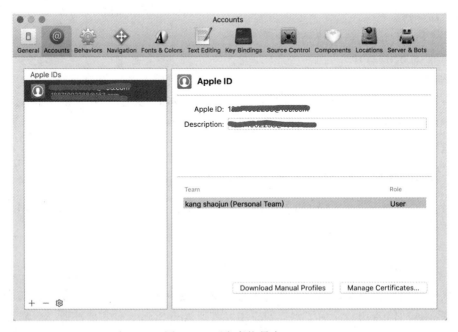

图 15-6　开发者帐号窗口

步骤 2：接着注册在弹出的窗口里选择 Apple ID 一项，点击 Continus。如图 15-7 所示。

图 15-7　选择创建帐号类型

步骤 3：接着注册一个 Apple ID 帐号，如果已经有，输入帐号及密码点击 Sign In 即可。如图 15-8 所示。

图 15-8　注册 Apple ID 窗口

 提示　App IDs 地址为：https://developer.apple.com/account/ios/identifier/bundle。

15.3.4　在 iTunes Connect 上创建应用程序记录

接下来，你将在 iTunes Connect 上注册你的应用程序：
- 在浏览器中打开 iTunes Connect。
- 在 iTunes Connect 登录页上，点击 My Apps。
- 点击 My App 页面左上角的 +，然后选择 New App。
- 填写你的应用详细信息。在 Platforms 部分中，确保已选中 iOS。由于 Flutter 目前不支持 tvOS，请不要选中该复选框。点击 Create。
- 导航到你 App 的应用程序详细信息 App Information 。
- 在 General Information 部分，选择你在上一步中注册的软件包 ID。

> 提示 要了解详细内容请参阅在 iTunes Connect 创建应用记录。请访问如下地址：https://developer.apple.com/library/content/documentation/LanguagesUtilities/Conceptual/iTunesConnect_Guide/Chapters/CreatingiTunesConnectRecord.html

15.3.5 查看 Xcode 项目设置

在 Xcode 中导航到你的 target 进行详细设置，具体步骤如下。

步骤 1：打开工程选中项目，如图 15-9 如所示。详细说明如下：

- 在 Xcode 中，在 helloworld/ios 文件夹下打开 Runner.xcworkspace。
- 要查看你的应用程序的设置，请在 Xcode 项目导航器中选择 Runner 项目。然后，在主视图边栏中选择 Runner Target。
- 选择 General 选项卡。

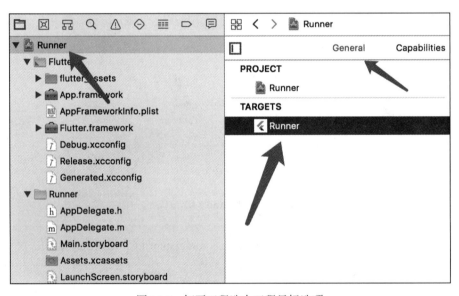

图 15-9　打开工程选中工程目标选项

步骤 2：验证最重要的设置。项目设置的 General 面板如图 15-10 所示。

在 Identity 部分：

- Display Name：应用程序的名称，例如：helloworld。
- Bundle Identifier：这是你在 iTunes Connect 上注册的 App ID。要保证唯一，例如：demo.com.helloworld。

在 Signing 部分：

- Automatically manage signing：Xcode 是否应该自动管理应用程序签名和生成。默认设置为 true。

❑ Team：选择开发者帐户关联的团队。如果需要，可以再添加帐号，然后更新此设置。

在 Deployment Info 部分：

❑ Deployment Target：将支持的最低 iOS 版本。Flutter 支持 iOS 8.0 及更高版本。有的应用可能需要更高版本的配置，请根据实际需要提高最低版本。

本项目设置的 General 选项卡应该类似于图 15-10 所示内容。

图 15-10　项目设置 General 面板

15.3.6　添加应用程序图标

iOS 应用同样需要精美的图标来展示你的应用，具体步骤如下：

1）查看 iOS App Icon 指南，设计程序图标。

2）在 Xcode 项目导航器中，在 Runner 文件夹中选择 Assets.xcassets。使用你自己的应用程序图标更换占位图标。如图 15-11 所示。你会发现你需要创建很多图标，才能满足实际的应用需要。

应用图标所使用的名称，请查看官方示例的工程图标，并点击右键查看其目录里的文件名。如图 15-12 所示。

296 ❖ Flutter 技术入门与实战

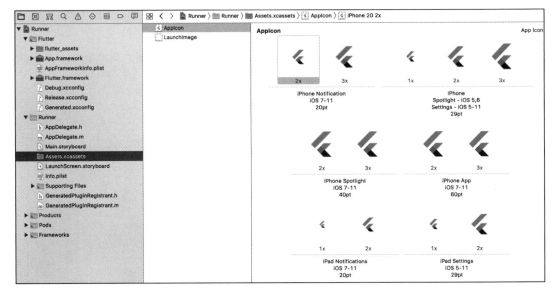

图 15-11　更换占位图标

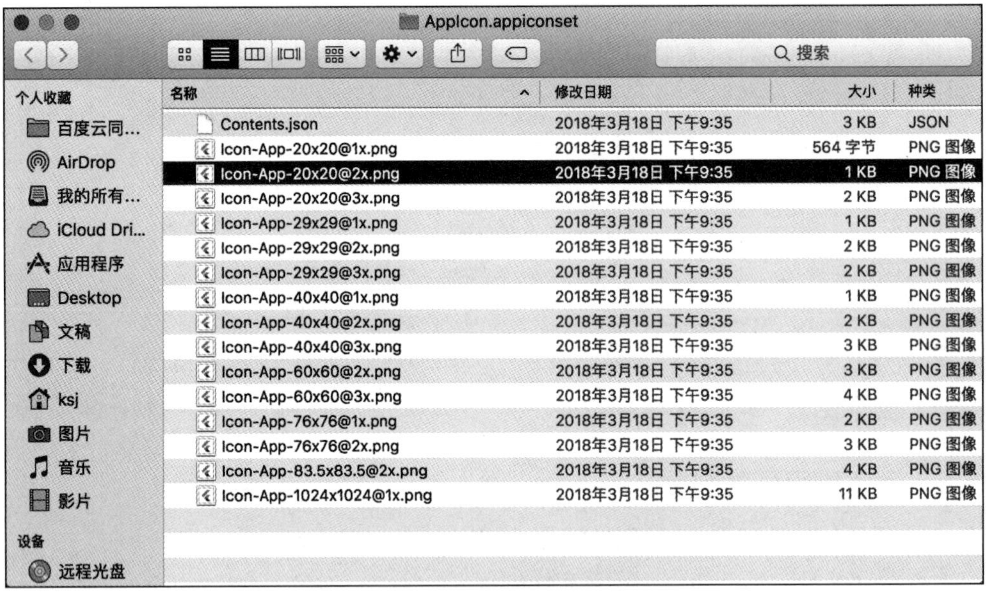

图 15-12　iOS 应用图标名称

3）运行 flutter run 验证应用图标已被替换。图标展示的效果如图 15-13 所示。请注意检查图标样式及工程名称是否正确。

图 15-13　iOS 应用程序图标

 提示　iOS App Icon 指南地址为：https://developer.apple.com/ios/human-interface-guidelines/graphics/app-icon/

15.3.7　准备发布版本

在这一步中，将创建一个构建档案并上传到 iTunes Connect。在开发过程中，应一直在构建、调试、测试 debug 版本。当你准备将应用发布到 App Store 上时，需要准备发布版本。在命令行里运行 flutter build ios 创建发布版本（flutter build 默认为 --release）。

在 Xcode 中，配置应用程序版本并构建，步骤如下：

1）在 Xcode 中，在你工程目录下的 ios 文件夹中打开 Runner.xcworkspace。选择 Product → Scheme → Edit Scheme，如图 15-14 所示。

2）把 Edit Scheme 容器左侧的

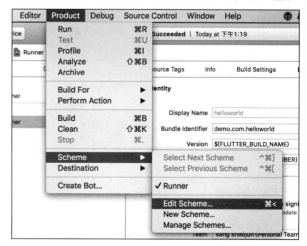

图 15-14　打开 Edit Scheme 菜单

选项都设置成 Release，如图 15-15 所示。

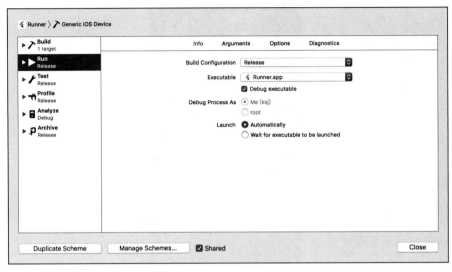

图 15-15　设置发布版

3）在 Xcode 项目导航器中选择 Runner，然后在设置视图边栏中选择 Runner target。在 Identity 部分中，将 Version 更新到你希望发布的版本号，如图 15-16 所示。

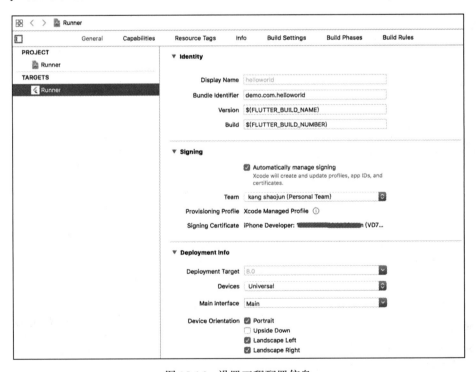

图 15-16　设置工程配置信息

4）选择 Product → Archive 以生成构建档案，如图 15-17 所示。

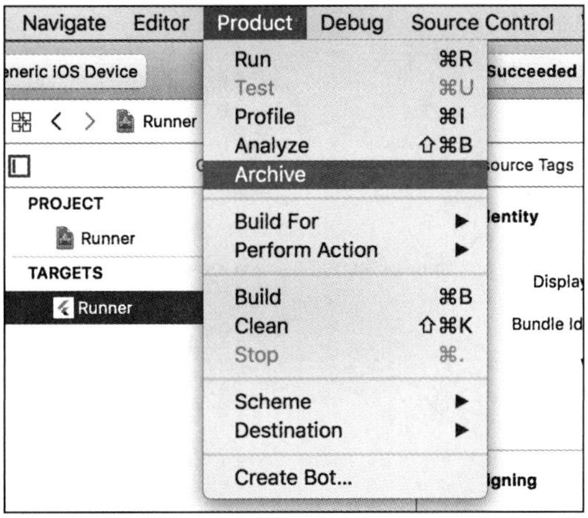

图 15-17　打包应用菜单

5）当 Xcode 完成归档后，点击 Upload to App Store… 按钮，选择开发团队，最后点击 Upload，等几分钟，如图 15-18 所示。

图 15-18　打包等待图

6）完成后，你会看到成功信息面板，如图 15-19 所示。

你应该在 30 分钟内收到一封电子邮件，通知你构建已经过验证，并可以在 TestFlight 上发布给测试人员。此时，你可以选择是否在 TestFlight 上发布，或继续将你的发布版发布到 App Store。

> 提示　TestFlight Beta 测试是苹果公司的产品，旨在更容易地邀请用户，在你发布产品到 App Store 之前，让他们能够参与测试你的 iOS、watchOS 和 tvOS 应用。

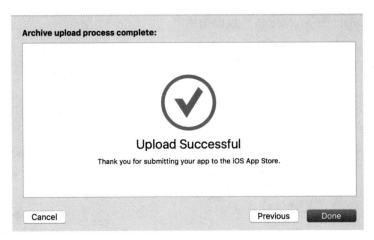

图 15-19 打包成功窗口

15.3.8 将应用发布到 App Store

当你准备将应用发布到全世界时，请按照以下步骤将你的应用提交给 App Store 进行审查和发布：

- ❏ 从 iTunes 应用程序的"应用程序详情页"的边栏中选择 Pricing and Availability，然后填写所需的信息。
- ❏ 从边栏选择状态。如果这是该应用的第一个版本，则其状态将为 1.0 Prepare for Submission。完成所有必填字段。
- ❏ 点击 Submit for Review。

Apple 公司会在应用程序审查过程完成时通知你。你的应用将根据你在 Version Release 部分指定的说明进行发布。

> 提示　要了解更多信息请参阅 Submitting Your App to the Store，访问地址为：https://developer.apple.com/library/content/documentation/IDEs/Conceptual/AppDistributionGuide/SubmittingYourApp/SubmittingYourApp.html。

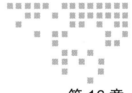

第 16 章 Chapter 16

综合案例——即时通讯 App 界面实现

到目前为止，我们学习了 Flutter 的基础组件、页面布局、手势处理及路由导航等知识。本章带领大家使用 Flutter 实现一个案例——即时通讯 App 的界面实现。

本章按以下几个方面来开发此应用：

- 项目介绍
- 项目搭建
- 入口程序
- 加载页面
- 应用页面
- 搜索页面
- 聊天页面
- 好友页面
- 我的页面

16.1 项目介绍

即时通讯软件是实现在线聊天、交流的软件，典型的代表有：QQ、百度 HI、Skype、Gtalk、新浪 UC、MSN 等。通常都包含以下几大功能：

- 文字聊天
- 群组聊天
- 好友管理

❏ 用户列表

Flutter 即时通讯 App 界面实现不包含服务端程序。采用 Flutter 最新版本 Flutter Release1.0（Flutter 第一个正式版本）。整体采用 Material Design 风格设计，构建一套精美的皮肤。项目中使用的插件有：

❏ flutter_webview_plugin：移动端浏览网页的插件

❏ date_format：日期格式化插件

应用的效果如图 16-1 所示。

图 16-1　App 界面效果图

16.2　项目搭建

本节带领大家一步步地把项目的框架搭建起来，包括添加资源、插件等内容。

16.2.1　新建项目

新建项目是项目开发的第一步，包括工程创建、资源添加、库添加、项目配置等内容。具体步骤如下：

步骤 1：新建工程请参考 1.3 节。基本的步骤是一样的，只是注意项目命名为 flutter_

im。另外再填写一个项目描述即可。

步骤 2：添加目录结构。请按图 16-2 所示的目录结构添加好内容。添加 images 目录，删除 test 目录。

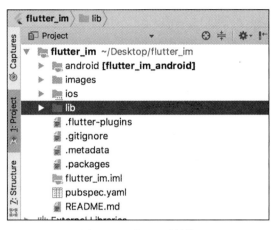

图 16-2　项目目录结构

步骤 3：准备好项目中使用的各种图标、背影图片、加载图片等位图资源。图片放入 images 目录下即可，如图 16-3 所示。

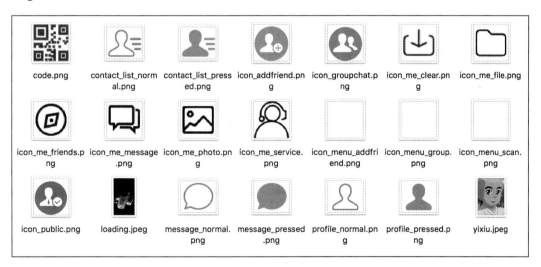

图 16-3　项目位图资源图

> **提示**：图标可以从图标库上下载：http://www.iconfont.cn。选择 48×48 大小的图片。

打开项目配置文件 pubspec.yaml，如图 16-4 所示。

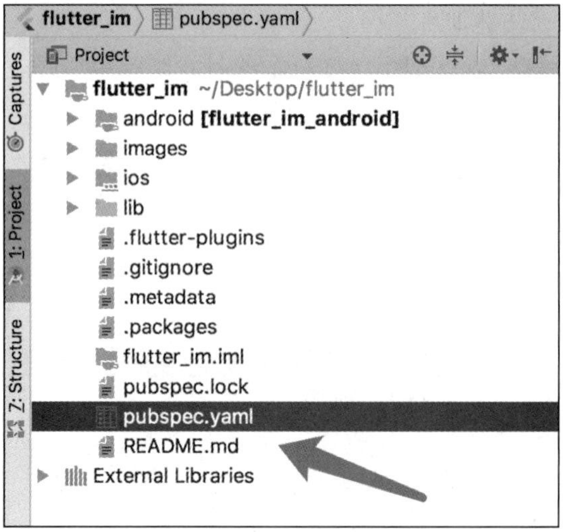

图 16-4　项目配置文件路径

在 assets 资源节点下添加项目中用到的所有位图资源。代码如下所示：

```
flutter:
  uses-material-design: true

  assets:

  # 加载页面图片资源
  - images/loading.jpeg

  # 底部 tab 切换图片资源
  - images/contact_list_normal.png
  - images/contact_list_pressed.png
  - images/profile_normal.png
  - images/profile_pressed.png
  - images/message_normal.png
  - images/message_pressed.png

  # 菜单按钮
  - images/icon_menu_addfriend.png
  - images/icon_menu_group.png

  # 好友模块图标
  - images/icon_addfriend.png
  - images/icon_groupchat.png
  - images/icon_public.png

  # 我的模块图标
  - images/icon_me_friends.png
  - images/icon_me_photo.png
  - images/icon_me_file.png
```

```
    - images/icon_me_service.png
    - images/icon_me_message.png
    - images/icon_me_clear.png
    - images/code.png
    - images/yixiu.jpeg
```

提
示　图片的配置要与 images 文件夹下的文件名保持一致。命名名遵循一定的规则。比如：icon_menu 就表示菜单图标

位图资源加完后，点击配置文件右上角"Packages get"更新配置。

步骤 4：再次打开 pubspec.yaml 文件，添加项目所需插件 flutter_webview_plugin 及 date_format。代码如下所示：

```
dependencies:
  flutter:
    sdk: flutter

  flutter_webview_plugin: "^0.3.0+2"
  date_format: "^1.0.4"
```

添加好配置后，一定要点击配置文件右上角"Packages get"，用来获取指定版本的插件。

16.2.2　添加源码目录及文件

新建好项目、配置好资源后，接下来就是编码工作。项目的源码位于 lib 目录下，包含聊天模块、好友模块、我的模块、加载页面、搜索页面、主页面、公共类及主程序。请按 lib 目录添加好子目录及源码文件，如下所示。

```
├── README.md(项目说明)
├── flutter_im.iml
├── lib(源码目录)
│   ├── app.dart(主页面)
│   ├── chat(聊天模块)
│   │   ├── message_data.dart(消息数据)
│   │   ├── message_item.dart(消息项)
│   │   └── message_page.dart(消息页面)
│   ├── common(公共类)
│   │   ├── im_item.dart(IM列表项)
│   │   └── touch_callback.dart(触摸回调封装)
│   ├── contacts(好友模块)
│   │   ├── contact_header.dart(好友列表头部)
│   │   ├── contact_item.dart(好友列表项)
│   │   ├── contact_sider_list.dart(好友列表)
│   │   ├── contact_vo.dart(好友vo类)
│   │   └── contacts.dart(好友页面)
│   ├── loading.dart(加载页面)
```

```
|       ├── main.dart（主程序）
|       ├── personal（我的模块）
|       |    └── personal.dart（我的页面）
|       └── search.dart（搜索页面）
├── pubspec.lock
└── pubspec.yaml（项目配置文件）
```

> 注意：项目结构文件是不包含 android、ios、images 及 build 目录的。

16.3 入口程序

所有的应用都有一个入口程序，通常是 main 函数引导进入应用程序。入口程序主要做以下几方面的处理：

- 自定义主题
- 定义路由表
- 指定首页

入口程序实现步骤如下：

步骤 1：通过自定义将主题定义为绿色小清新风格。定义导航栏、弹出菜单为绿色。如下代码所示：

```
primaryColor: Colors.green
```

步骤 2：定义路由表为整个应用程序作导航使用。整个应用分三块需要路由，分别是：

- 应用程序：app
- 好友动态：friends
- 搜索页面：search

这里要重点说一下 WebviewScaffold 组件，WebviewScaffold 是 Flutter 用来实现网页浏览的组件，非常实用。项目里我们用来模拟好友动态页面，大致用法如下所示：

```
"/friends": (_) => new WebviewScaffold(
  url: "指定网址 url",
  appBar: new AppBar(
    title: // 指定打开网页的标题
  ),
  withZoom://是否缩放页面
  withLocalStorage://是否本地存储
),
```

步骤 3：打开应用的第一个页面即为首页。通常首页并不是应用程序的主画面，而是加载页面，指定方式如下所示：

```
home: new LoadingPage()
```

步骤 4：打开 lib/main.dart 文件，按照上面几个步骤编写如下完整代码：

```
import 'package:flutter/material.dart';
import './app.dart';
import './loading.dart';
import 'package:flutter_webview_plugin/flutter_webview_plugin.dart';
import './search.dart';

// 应用程序入口
void main() => runApp(MaterialApp(
      debugShowCheckedModeBanner: false,
      title: '聊天室',
      // 自定义主题
      theme: mDefaultTheme,
      // 添加路由表
      routes: <String, WidgetBuilder>{
        "app": (BuildContext context) => new App(),
        "/friends": (_) => new WebviewScaffold(
            //Webview 插件
            url: "https://flutter.io/",
            appBar: new AppBar(
              title: new Text("Flutter 官网"),
            ),
            withZoom: true,
            withLocalStorage: true,
          ),
        'search': (BuildContext context) => new Search(), // 搜索页面路由
      },
      // 指定首页，默认为加载页面
      home: new LoadingPage(),
    ));

// 自定义主题 绿色小清新风格
final ThemeData mDefaultTheme = new ThemeData(
  primaryColor: Colors.green,
  scaffoldBackgroundColor: Color(0xFFebebeb),
  cardColor: Colors.green,
);
```

16.4 加载页面

其实加载页面和普通的页面并没有什么两样，唯一的区别是，加载页面是伴随着应用程序的加载完成的。比如：以即时通讯 App 为例，应用程序需要在此阶段完成连接 socket、下载资源文件、请求广告信息等。由于这个过程是需要时间处理的，所以这个页面需要停留一定的时间，通常设置成几秒即可。代码如下所示：

```
Future.delayed(Duration(seconds: 3),(){
  //TODO
}
```

上面的代码使用 Future.delayed 函数，使得加载页面停留 3 秒，3 秒过后页面马上要跳转到应用程序页面，否则应用将永远停留在加载页面。应用跳转采用 Navigator 导航器实现，代码如下所示：

```
Navigator.of(context).pushReplacementNamed("app");
```

加载页面通常为广告页面及产品介绍页面。比如用户首次使用 App 则要引导用户如何使用主要功能。在这里我们用一张漂亮的背景图片作为加载页面的全部展示内容。这里我们需要打开 lib/loading.dart 文件，添加下面的所有代码：

```
import 'package:flutter/material.dart';
import 'dart:async';

// 加载页面
class LoadingPage extends StatefulWidget {
  @override
  _LoadingState createState() => new _LoadingState();
}

class _LoadingState extends State<LoadingPage> {

  @override
  void initState(){
    super.initState();
    // 在加载页面停顿 3 秒
    new Future.delayed(Duration(seconds: 3),(){
      print("Flutter 即时通讯 APP 界面实现....");
      Navigator.of(context).pushReplacementNamed("app");
    });
  }

  @override
  Widget build(BuildContext context) {
    return new Center(
      child: Stack(
        children: <Widget>[
          // 加载页面居中背景图 使用 cover 模式
          Image.asset("images/loading.jpeg",fit: BoxFit.cover,),
        ],
      ),
    );
  }

}
```

> **注意**　加载页面停顿要放在 initState 函数里处理，原因是必须等待页面渲染完成才行，否则加载的画面内容就看不到了。

加载页面的效果如图 16-5 所示。

图 16-5　加载页面效果

16.5　应用页面

应用页面为加载页面跳转后进入的页面。文件路径为 lib/app.dart。应用页面为整个 App 的核心页面，包括顶部导航栏、右上角功能菜单、中间功能模块、底部导航按钮。先看应用页面的整体效果，如图 16-6 所示。

接下来我们一步步实现它。

步骤 1：创建应用页面类 App，继承自 StatefulWidget。由于页面里有切换的状态变换，所以必需要继承有状态的 Widget。同时要添加状态实现类 AppState。继承代码如下所示：

```
class App extends StatefulWidget
```

步骤 2：导入各个功能页面，当然目前还没有实现。提前写上后面逐步补充完整。需要导入的页面有以下几个：

❑ MessagePage：聊天页面

❑ Contacts：好友页面
❑ Personal：我的页面

由于搜索页面属于跳转页所以这里不需要导入。导入好页面后，需要写一个方法根据当前索引返回不同的页面，代码结构如下所示：

```
currentPage() {
  switch (_currentIndex) {
    case 0:
      // 返回聊天页面
    case 1:
      // 返回好友页面
    case 2:
      // 返回我的页面
    default:
  }
}
```

步骤3：编写弹出菜单项方法，用PopupMenuItem组件构建。菜单项方法需要传入两个参数菜单标题、图片路径或图标。注意图片路径和图标参数是二选一，一次只能传入一个。方法定义如下所示：

```
_popupMenuItem(String title, {String imagePath, IconData icon})
```

图16-6　应用页面预览

菜单项的展示是左右结构的，所以这里需要使用Row组件包装。图片路径及图标需要做一个三元表达式的判断，代码结构如下所示：

```
imagePath != null
  ? Image.asset(
    imagePath,
    )
  : SizedBox(
    icon,
    ),
```

步骤4：使用Scaffold组件来渲染整个页面。包含以下三大块：
❑ appBar：顶部导航栏，包含标题、搜索及菜单按钮。
❑ bottomNavigationBar：底部导航按钮，包含"聊天"、"好友"、"我的"三个按钮。
❑ body：中间功能模块内容，包含聊天模块、好友模块、我的模块。

步骤5：编写标题、搜索及菜单按钮。其中搜索按钮点击需要路由到搜索页面。代码如下所示：

```
Navigator.pushNamed(context, 'search');
```

点击"菜单"按钮，需要调用 showMenu 方法来弹出菜单列表，调用示例如下所示：

```
showMenu(
    context: context,
    items: <PopupMenuEntry>[
      _popupMenuItem('发起会话',imagePath: 'images/icon_menu_group.png'),
      // 其他菜单
      ],
);
```

由于弹出菜单在界面的右上角，所以这里还需要给菜单一个定位处理，使用 RelativeRect.fromLTRB 函数，取距离边框左上右下值。处理代码如下所示：

```
position: RelativeRect.fromLTRB(500.0, 76.0, 10.0, 0.0),
```

> **注意** 底部取值为 0.0。表示底部值不考虑。

编写完标题、搜索及菜单按钮后，将组装好的组件赋值给 Scaffold 组件的 appBar 属性即可。

步骤 6：底部导航按钮采用 BottomNavigationBar 组件实现，每一个按钮使用的组件是 BottomNavigationBarItem 来实现。实现的关键点在于选中及没有选中按钮图标及颜色的控制。以聊天按钮实现为例，采用三元表达式判断当前选中按钮的索引，代码结构如下所示：

```
BottomNavigationBarItem(
  title: new Text(
      '聊天',
      style: TextStyle(
        color: _currentIndex == 0 ? 选中颜色 : 未选中颜色,
      ),
   icon: _currentIndex == 0 ? 选中图标 : 未选中图标
```

BottomNavigationBar 还需要添加 onTap 回调方法，回调方法里添加当前选中按钮索引设置处理，代码如下所示：

```
onTap: ((index) {
      setState(() {
        _currentIndex = index;
      });
}
```

编写完底部导航按钮后，将组装好的组件赋值给 Scaffold 组件的 bottomNavigationBar 属性即可。

步骤 7：调用前面编写好的 currentPage 函数，将其返回对象赋值给 Scaffold 组件的 body 属性即可。

将前面所有步骤的代码串起来完整代码如下所示：

```dart
import 'package:flutter/material.dart';
import './chat/message_page.dart';
import './contacts/contacts.dart';
import './personal/personal.dart';

// 应用页面使用有状态 Widget
class App extends StatefulWidget {
  @override
  AppState createState() => AppState();
}

// 应用页面状态实现类
class AppState extends State<App> {
  // 当前选中页面索引
  var _currentIndex = 0;

  // 聊天页面
  MessagePage message;

  // 好友页面
  Contacts contacts;

  // 我的页面
  Personal me;

  // 根据当前索引返回不同的页面
  currentPage() {
    switch (_currentIndex) {
      case 0:
        if (message == null) {
          message = new MessagePage();
        }
        return message;
      case 1:
        if (contacts == null) {
          contacts = new Contacts();
        }
        return contacts;
      case 2:
        if (me == null) {
          me = new Personal();
        }
        return me;
      default:
    }
  }

  // 渲染某个菜单项，传入菜单标题，图片路径或图标
  _popupMenuItem(String title, {String imagePath, IconData icon}) {
    return PopupMenuItem(
```

```
      child: Row(
        children: <Widget>[
          // 判断是使用图片路径还是图标
          imagePath != null
              ? Image.asset(
                  imagePath,
                  width: 32.0,
                  height: 32.0,
                )
              : SizedBox(
                  width: 32.0,
                  height: 32.0,
                  child: Icon(
                    icon,
                    color: Colors.white,
                  ),
                ),
          // 显示菜单项文本内容
          Container(
            padding: const EdgeInsets.only(left: 20.0),
            child: Text(
              title,
              style: TextStyle(color: Colors.white),
            ),
          ),
        ],
      ),
    );
}

@override
Widget build(BuildContext context) {
  return Scaffold(
    appBar: AppBar(
      title: Text(' 即时通讯 '),
      actions: <Widget>[
        GestureDetector(
          onTap: () {
            // 跳转至搜索页面
            Navigator.pushNamed(context, 'search');
          },
          child: Icon(
            // 搜索图标
            Icons.search,
          ),
        ),
        Padding(
          // 左右内边距
          padding: const EdgeInsets.only(left: 30.0, right: 20.0),
          child: GestureDetector(
```

```dart
        onTap: () {
          // 弹出菜单
          showMenu(
            context: context,
            // 定位在界面的右上角
            position: RelativeRect.fromLTRB(500.0, 76.0, 10.0, 0.0),
            // 展示所有菜单项
            items: <PopupMenuEntry>[
              _popupMenuItem('发起会话',
                  imagePath: 'images/icon_menu_group.png'),
              _popupMenuItem('添加好友',
                  imagePath: 'images/icon_menu_addfriend.png'),
              _popupMenuItem('联系客服', icon: Icons.person),
            ],
          );
        },
        // 菜单按钮
        child: Icon(Icons.add),
      ),
    ),
  ],
),
// 底部导航按钮
bottomNavigationBar: new BottomNavigationBar(
  // 通过 fixedColor 设置选中 item 的颜色
  type: BottomNavigationBarType.fixed,
  // 当前页面索引
  currentIndex: _currentIndex,
  // 按下后设置当前页面索引
  onTap: ((index) {
    setState(() {
      _currentIndex = index;
    });
  }),
  // 底部导航按钮项
  items: [
    // 导航按钮项传入文本及图标
    new BottomNavigationBarItem(
      title: new Text(
        '聊天',
        style: TextStyle(
            color: _currentIndex == 0
                ? Color(0xFF46c01b)
                : Color(0xff999999)),
      ),
      // 判断当前索引作图片切换显示
      icon: _currentIndex == 0
          ? Image.asset(
              'images/message_pressed.png',
              width: 32.0,
              height: 28.0,
```

```
                )
            : Image.asset(
                'images/message_normal.png',
                width: 32.0,
                height: 28.0,
              )),
          new BottomNavigationBarItem(
              title: new Text(
                '好友',
                style: TextStyle(
                    color: _currentIndex == 1
                        ? Color(0xFF46c01b)
                        : Color(0xff999999)),
              ),
              icon: _currentIndex == 1
                  ? Image.asset(
                      'images/contact_list_pressed.png',
                      width: 32.0,
                      height: 28.0,
                    )
                  : Image.asset(
                      'images/contact_list_normal.png',
                      width: 32.0,
                      height: 28.0,
                    )),
          new BottomNavigationBarItem(
              title: new Text(
                '我的',
                style: TextStyle(
                    color: _currentIndex == 2
                        ? Color(0xFF46c01b)
                        : Color(0xff999999)),
              ),
              icon: _currentIndex == 2
                  ? Image.asset(
                      'images/profile_pressed.png',
                      width: 32.0,
                      height: 28.0,
                    )
                  : Image.asset(
                      'images/profile_normal.png',
                      width: 32.0,
                      height: 28.0,
                    )),
        ],
      ),
      // 中间显示当前页面
      body: currentPage(),
    );
  }
}
```

16.6 搜索页面

点击应用页面的"搜索"按钮就可以跳转到搜索页面。搜索页面包含上部导航部分及下面搜索分类两大部分。其中导航部分包含返回按钮、搜索框及麦克风按钮。搜索分类主要是两排常用的搜索项。

16.6.1 布局拆分

在编写页面前需要首先拆分页面，可以先画草图，仔细想好每部分大概要用到什么布局组件。接着就是思考细节问题，如输入框下划线怎么处理。搜索页面的布局拆分及效果预览如图 16-7 所示。首先把整个页面拆分成 4 个部分，垂直方向用一个 Column 包裹，最上面为导航栏。导航栏使用水平组件 Row 包裹，包含返回按钮、搜索框及麦克风按钮。接下来是搜索分类标题及搜索具体分类，分别用三行表示。

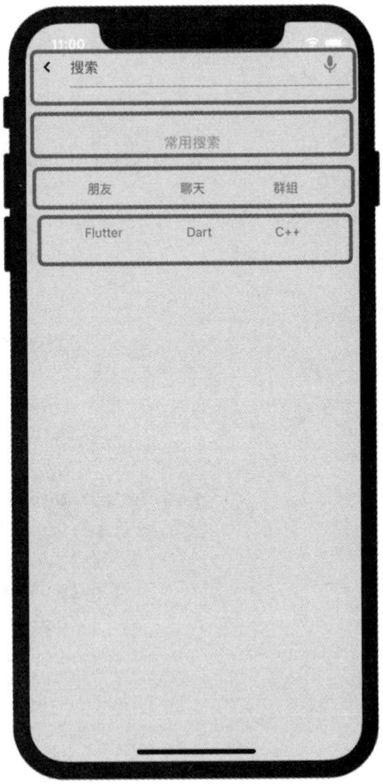

16.6.2 请求获取焦点

由于搜索页面需要接收用户的输入操作，为了让用户一打开这个界面就能马上输入搜索内容，所以需要在界面渲染完成后让搜索框获取好焦点。解决的办法如下所示。

定义一个焦点节点，使用 FocusNode，代码如下所示：

```
FocusNode focusNode = new FocusNode();
```

图 16-7 搜索页面布局拆分图

编写一个请求获取焦点的方法，把这个方法返回的对象赋值给搜索框的 focusNode 属性即可。请求方法代码如下所示：

```
requestFocus() {
  FocusScope.of(context).requestFocus(focusNode);
  return focusNode;
}
```

16.6.3 自定义 TouchCallBack 组件

TouchCallBack 组件是用来做触摸回调使用的。为什么自定义这个组件，主要是因为项目里多次会用到，主要是起到复用的作用。省得每个要用到的地方都写相同的代码，增加了代码的冗余量。

TouchCallBack 组件需要传入两个最重要的参数：
- @required this.child：子组件
- @required this.onPressed：回调函数

 注意 参数列表里加入了 @required 关键字表示此参数必须传。

另外，封装代码里需要添加 GestureDetector 组件，用来做触摸响应。完整的实现代码如下所示。这段代码是放在 lib/common/ touch_callback.dart 文件里的。其他模块也需要用到。

```dart
import 'package:flutter/material.dart';

// 触摸回调组件
class TouchCallBack extends StatefulWidget{
  // 子组件
  final Widget child;
  // 回调函数
  final VoidCallback onPressed;
  final bool isfeed;
  // 背景色
  final Color background;
  // 传入参数列表
  TouchCallBack({Key key,
    @required this.child,
    @required this.onPressed,
    this.isfeed:true,
    this.background:const Color(0xffd8d8d8),
  }):super(key:key);
  @override
  TouchState createState() => TouchState();
}

class TouchState extends State<TouchCallBack>{
  Color color =  Colors.transparent;
  @override
  Widget build(BuildContext context) {
    // 返回 GestureDetector 对象
    return GestureDetector(
      // 使用 Container 容器包裹
      child: Container(
        color: color,
        child: widget.child,
      ),
      //onTap 回调
      onTap: widget.onPressed,
      onPanDown: (d){
        if(widget.isfeed == false) return;
        setState((){
```

```
        color = widget.background;
      });
    },
    onPanCancel: (){
      setState(() {
        color = Colors.transparent;
      });
    },
  );
}
```

16.6.4 返回文本组件

由于搜索页面里需要多个搜索分类文本显示,所以最好也封装一个返回文本的方法。代码如下所示:

```
_getText(String text) {
  // 返回 Text
}
```

16.6.5 组装实现搜索页面

理清了布局的思路,封装好复用的组件,处理好关键代码。剩下的就是组装实现各个部分了,最终完成搜索页面的实现。打开 lib/search.dart 编写如下完整代码:

```
import 'package:flutter/material.dart';
import './common/touch_callback.dart';

// 搜索模块
class Search extends StatefulWidget {
  @override
  SearchState createState() => new SearchState();
}

class SearchState extends State<Search> {
  // 定义焦点节点
  FocusNode focusNode = new FocusNode();

  // 请求获取焦点
  _requestFocus() {
    FocusScope.of(context).requestFocus(focusNode);
    return focusNode;
  }

  // 返回一个文本组件
  _getText(String text) {
```

```
    return TouchCallBack(
      isfeed: false,
      onPressed: () {},
      child: Text(
        text,
        // 添加文本样式
        style: TextStyle(fontSize: 14.0, color: Color(0xff1aad19)),
      ),
    );
}

// 搜索页面渲染
@override
Widget build(BuildContext context) {
  return Scaffold(
    body: Container(
      // 顶部留一定距离
      margin: const EdgeInsets.only(top: 25.0),
      // 整体垂直布局
      child: Column(
        // 水平方向居中
        crossAxisAlignment: CrossAxisAlignment.center,
        children: <Widget>[
          // 顶部导航栏包括返回按钮、搜索框及麦克风按钮
          Stack(
            children: <Widget>[
              // 使用触摸回调组件
              TouchCallBack(
                isfeed: false,
                onPressed: () {
                  // 使用导航器返回上一个页面
                  Navigator.pop(context);
                },
                child: Container(
                  height: 45.0,
                  margin: const EdgeInsets.only(left: 12.0, right: 10.0),
                  // 添加返回按钮
                  child: Icon(
                    Icons.chevron_left,
                    color: Colors.black,
                  ),
                ),
              ),
              // 搜索框容器
              Container(
                alignment: Alignment.centerLeft,
                height: 45.0,
                margin: const EdgeInsets.only(left: 50.0, right: 10.0),
                // 搜索框底部边框
                decoration: BoxDecoration(
```

```dart
          border: Border(
              bottom: BorderSide(width: 1.0, color: Colors.green)),
        ),
        child: Row(
          crossAxisAlignment: CrossAxisAlignment.center,
          children: <Widget>[
            Expanded(
              // 输入框
              child: TextField(
                // 请求获取焦点
                focusNode: _requestFocus(),
                style: TextStyle(
                  color: Colors.black,
                  fontSize: 16.0,
                ),
                onChanged: (String text) {},
                decoration: InputDecoration(
                    hintText: '搜索', border: InputBorder.none),
              ),
            ),
            // 添加麦克风图标
            Container(
              margin: const EdgeInsets.only(right: 10.0),
              child: Icon(
                Icons.mic,
                color: Color(0xffaaaaaa),
              ),
            ),
          ],
        ),
      ),
    ],
  ),
),
Container(
  margin: const EdgeInsets.only(top: 50.0),
  child: Text(
    '常用搜索',
    style: TextStyle(fontSize: 16.0, color: Color(0xffb5b5b5)),
  ),
),
Padding(
  padding: const EdgeInsets.all(30.0),
  child: Row(
    // 对齐方式采用均匀对齐
    mainAxisAlignment: MainAxisAlignment.spaceAround,
    // 第一行搜索项
    children: <Widget>[
      _getText('朋友'),
      _getText('聊天'),
      _getText('群组'),
```

```
              ],
            ),
          ),
          Padding(
            padding: const EdgeInsets.only(left: 30.0, right: 30.0),
            child: Row(
              mainAxisAlignment: MainAxisAlignment.spaceAround,
              children: <Widget>[
                //第二行搜索项
                _getText('Flutter'),
                _getText('Dart'),
                _getText('C++'),
              ],
            ),
          ),
        ],
      ),
    ),
  );
}
}
```

16.7 聊天页面

这里我们主要实现聊天信息列表展示功能。代码拆分成以下三个文件:

❑ message_data.dart：组织聊天数据。
❑ message_item.dart：聊天信息列表项。
❑ message_page.dart：聊天信息列表页。

首先预览一下聊天页面效果，如图 16-8 所示。

图 16-8 聊天页面效果

16.7.1 准备聊天消息数据

聊天消息数据相关代码都是放在 lib/chat/message_data.dart 文件里的，所以请将本节接下来所列代码放在这个文件里，步骤如下。

步骤 1：定义消息枚举类型，通常聊天消息有系统消息、公共消息、私聊消息、群聊消息。代码如下所示：

```
// 消息类型枚举类型
enum MessageType { SYSTEM,PUBLIC,CHAT,GROUP }
```

步骤 2：定义聊天数据 VO 类。包含头像、主标题、子标题、消息时间、消息类型字段。代码如下所示：

```
// 聊天数据
class MessageData{
  // 头像
  String avatar;
  // 主标题
  String title;
  // 子标题
  String subTitle;
  // 消息时间
  DateTime time;
  // 消息类型
  MessageType type;

  MessageData(this.avatar,this.title,this.subTitle,this.time,this.type);
}
```

步骤3：定义好了聊天数据VO类及消息类型，我们就可以组装数据了。使用列表 List<MessageData> 做数据存储。代码如下所示：

```
List<MessageData> messageData = [
  new
MessageData('https://timgsa.baidu.com/timg?image&quality=80&size=b9999_10000&sec
=1544070910437&di=86ffd13f433c252d4c49afe822e87462&imgtype=0&src=http%3A%2F%2Fim
gsrc.baidu.com%2Fforum%2Fw%3D580%2Fsign%3Debf3e26b1a4c510faec4e21250582528%2F0cf
431adcbef76092781a53c2edda3cc7dd99e8e.jpg',
    '一休哥',
    '突然想到的',
    new DateTime.now(),
    MessageType.CHAT
  ),
  new MessageData(
'https://timgsa.baidu.com/timg?image&quality=80&size=b9999_10000&sec=15404032826
49&di=c4f237332e6bf94546c950817699c2fd&imgtype=0&src=http%3A%2F%2Fimg5q.duitang.
com%2Fuploads%2Fitem%2F201504%2F11%2F20150411H0128_PHr4z.jpeg',
    '多拉a梦',
    '机器猫！！！',
    new DateTime.now(),
    MessageType.CHAT
  ),
  // 添加更多聊天消息数据
];
```

 为了让聊天列表显示得数据更丰富些，建议多加一些MessageData。

16.7.2 聊天消息列表项实现

这里我们先展示一条聊天消息，需要编写一个MessageItem组件。由于组件是纯展示界面的，所以只需要继承 StatelessWidget 即可。消息列表里需要展示消息时间，这里需要对时间进行格式化处理，所以需要导入 date_format 插件，代码如下所示：

```
import 'package:date_format/date_format.dart';
```

越是小的模块，越是需要拆分得很细。如图 16-9 所示，我们拿三条聊天消息用来做布局拆分分析。整体采用水平布局，最左侧为头像，中间是主标题及子标题，最右侧是时间。

图 16-9　消息列表项布局拆分图

有了上面的分析，我们看一下聊天消息列表项的完整代码，打开 lib/chat/message_item.dart 文件，编写如下代码：

```
import 'package:flutter/material.dart';
import './message_data.dart';
import 'package:date_format/date_format.dart';
import '../common/touch_callback.dart';

//聊天信息项
class MessageItem extends StatelessWidget{
  final MessageData message;
  MessageItem(this.message);
  @override
  Widget build(BuildContext context) {
    //最外层容器
    return Container(
      decoration: BoxDecoration(
        color: Colors.white,
        //仅加一个底部边，这样整个列表的每项信息下面都会有一条边
        border: Border(bottom: BorderSide(width: 0.5,color: Color(0xFFd9d9d9))),
      ),
      height: 64.0,
      //按下回调处理空实现
      child: TouchCallBack(
        onPressed: (){
        },
        //整体水平方向布局
        child: Row(
          //垂直方向居中显示
          crossAxisAlignment: CrossAxisAlignment.center,
```

```dart
      children: <Widget>[
        // 展示头像
        Container(
          // 头像左右留一定的外边距
          margin: const EdgeInsets.only(left: 13.0,right: 13.0),
          child: Image.network(message.avatar,width: 48.0,height: 48.0,),
        ),
        Expanded(
          // 主标题和子标题采用垂直布局
          child: Column(
            // 垂直方向居中对齐
            mainAxisAlignment: MainAxisAlignment.center,
            // 水平方向靠左对齐
            crossAxisAlignment: CrossAxisAlignment.start,
            children: <Widget>[
              Text(
                message.title,
                style: TextStyle(fontSize: 16.0,color: Color(0xFF353535)),
                maxLines: 1,
              ),
              Padding(
                padding: const EdgeInsets.only(top:8.0),
              ),
              Text(
                message.subTitle,
                style: TextStyle(fontSize: 14.0,color: Color(0xFFa9a9a9)),
                maxLines: 1,
                // 显示不完的文本用省略号来表示
                overflow: TextOverflow.ellipsis,
              ),
            ],
          ),
        ),
        Container(
          // 时间顶部对齐
          alignment: AlignmentDirectional.topStart,
          margin: const EdgeInsets.only(right: 12.0,top: 12.0),
          child: Text(
            // 格式化时间
            formatDate(message.time, [HH,':',nn,':','ss']).toString(),
            style: TextStyle(fontSize: 14.0,color: Color(0xFFa9a9a9)),
          ),
        ),
      ],
    ),
   ),
  );
 }
}
```

16.7.3 聊天消息列表实现

有了聊天消息数据及聊天消息列表项组件，只需要使用 ListView.builder 构造列表即可。打开 lib/chat/message_page.dart 文件，编写如下代码：

```
import 'package:flutter/material.dart';
import './message_data.dart';
import './message_item.dart';

// 聊天主页面
class MessagePage extends StatefulWidget{
  @override
  MessagePageState createState() => new MessagePageState();
}

class MessagePageState extends State<MessagePage>{
  @override
  Widget build(BuildContext context) {
    return Scaffold(
      // 构造列表
      body: ListView.builder(
        // 传入数据长度
        itemCount: messageData.length,
        // 构造列表项
        itemBuilder: (BuildContext context, int index){
          // 传入 messageData 返回列表项
          return new MessageItem(messageData[index]);
        }
      ),
    );
  }
}
```

16.8 好友页面

这里我们主要实现好友列表展示功能。代码拆分成以下几个文件：
- contact_header.dart：好友头部内容。
- contact_item.dart：好友列表项。
- contact_sider_list.dart：好友列表渲染。
- contact_vo.dart：好友列表数据。
- contacts.dart：好友列表页。

首先，预览一下好友页面效果，如图 16-10 所示。其中"新加好友""公共聊天室"为好友列表头部信息。字母 A、B、C……将好友数据按字母顺序排列并归类。剩下的就是好友列表项了，包含头像及名称。

图 16-10　好友列表效果图

16.8.1　准备好友列表数据

好友列表数据相关代码都放在 lib/contacts/contact_vo.dart 文件中，所以请将接下来的两步代码编写进这个文件里，步骤如下所示：

步骤 1：定义好友数据 VO 类。类命名为 ContactVO。包含字母排列值、名称、头像 url 字段。代码如下所示：

```
class ContactVO {
  // 字母排列值
  String seationKey;
  // 名称
  String name;
  // 头像 url
  String avatarUrl;
  // 构造函数
  ContactVO({@required this.seationKey,this.name,this.avatarUrl});
}
```

步骤 2：定义好了好友数据 VO 类，我们就可以组装数据了。使用列表 List<ContactVO> 做数据存储。代码如下所示：

```
List<ContactVO> contactData = [
  new ContactVO(
    seationKey: 'A',
    name: 'A 张三',
    avatarUrl:
'https://timgsa.baidu.com/timg?image&quality=80&size=b9999_10000&sec=1540633615163&di=e6662df8230b7e8a87cf0017df7252b7&imgtype=0&src=http%3A%2F%2Fimgsrc.baidu.com%2Fimgad%2Fpic%2Fitem%2Fac345982b2b7d0a2d1f5b989c0ef76094b369ae2.jpg',
  ),
  new ContactVO(
    seationKey: 'A',
    name: ' 阿黄 ',
    avatarUrl:
'https://timgsa.baidu.com/timg?image&quality=80&size=b9999_10000&sec=1540441383345&di=39fe0512213122b232a7861363d86ba4&imgtype=0&src=http%3A%2F%2Fimgsrc.baidu.com%2Fimgad%2Fpic%2Fitem%2F8435e5dde71190ef0795a828c41b9d16fcfa60de.jpg'),
  new ContactVO(
    seationKey: 'B',
    name: ' 波波 ',
    avatarUrl:
'https://timgsa.baidu.com/timg?image&quality=80&size=b9999_10000&sec=1540441383345&di=41a9c62adb0702595cbeab1eb7935f66&imgtype=0&src=http%3A%2F%2Fimgsrc.baidu.com%2Fimgad%2Fpic%2Fitem%2F3b292df5e0fe992532fd5c7e3fa85edf8db1712e.jpg'),
  // 更多好友数据
];
```

提示　好友肯定不止这三条数据，你可以根据需要再加多一些数据。seationKey 是用来做列表排序使用的必填字段。

16.8.2　好友列表项实现

好友列表项的布局会比聊天消息列表项简单很多。唯一的区别是，好友列表项还有一个列表头需要处理。所以好友列表项多了标题名 titleName 及图片路径 imageName 两个字段。我们只需要在页面渲染时和 ContactVO 数据做一下区分即可。

页面布局不做具体分析，直接看实现代码。打开 lib/contacts/contact_item.dart 文件。编写如下代码：

```
import 'package:flutter/material.dart';
import './contact_vo.dart';

// 好友列表项
class ContactItem extends StatelessWidget {
  // 好友数据 VO
  final ContactVO item;
```

```dart
// 标题名
final String titleName;
// 图片路径
final String imageName;
// 构建方法
ContactItem({this.item, this.titleName, this.imageName});

// 渲染好友列表项
@override
Widget build(BuildContext context) {
  // 列表项容器
  return Container(
    decoration: BoxDecoration(
      color: Colors.white,
      // 每条列表项底部加一个边框
      border:
          Border(bottom: BorderSide(width: 0.5, color: Color(0xFFd9d9d9))),
    ),
    height: 52.0,
    child: FlatButton(
      onPressed: () {},
      child: Row(
        crossAxisAlignment: CrossAxisAlignment.center,
        children: <Widget>[
          // 展示头像或图片
          imageName == null
              ? Image.network(
                  item.avatarUrl != ''
                      ? item.avatarUrl
                      :
'https://timgsa.baidu.com/timg?image&quality=80&size=b9999_10000&sec=15440709104
37&di=86ffd13f433c252d4c49afe822e87462&imgtype=0&src=http%3A%2F%2Fimgsrc.baidu.
com%2Fforum%2Fw%3D580%2Fsign%3Debf3e26b1a4c510faec4e21250582528%2F0cf431adcbef76
092781a53c2edda3cc7dd99e8e.jpg',
                  width: 36.0,
                  height: 36.0,
                  scale: 0.9,
                )
              : Image.asset(
                  imageName,
                  width: 36.0,
                  height: 36.0,
                ),
          // 展示名称或标题
          Container(
            margin: const EdgeInsets.only(left: 12.0),
            child: Text(
              titleName == null ? item.name ?? '暂时 ':titleName,
              style: TextStyle(fontSize: 18.0,color: Color(0xFF353535)),
              maxLines: 1,
```

```
          ),
        ),
      ],
    ),
  ),
);
    }
}
```

16.8.3 好友列表头实现

好友列表头主要是添加"新加好友"及"公共聊天室"两项数据。打开 lib/contacts/contact_header.dart 文件，按如下代码编写即可：

```
import 'package:flutter/material.dart';
import './contact_item.dart';

// 好友列表头部
class ContactHeader extends StatelessWidget {
  @override
  Widget build(BuildContext context) {
    return Column(children: <Widget>[
      ContactItem(titleName:' 新加好友 ',imageName:'images/icon_addfriend.png'),
      ContactItem(titleName:' 公共聊天室 ',imageName:'images/icon_groupchat.png'),
    ],);
  }
}
```

16.8.4 ContactSiderList 类

ContactSiderList 类是好友列表的具体实现，此类需要传入三个构造器及列表数据：
- items：好友列表数据。
- headerBuilder：好友列表头构造器。
- itemBuilder：好友列表项构造器。
- sectionBuilder：字母构造器。

打开 lib/contacts/contact_sider_list.dart 文件，按以下步骤编写代码：

步骤 1：创建 ContactSiderList 类，需要继承自 StatefulWidget。添加列表数据及构造器，代码结构如下所示：

```
class ContactSiderList extends StatefulWidget {
    // 添加列表数据及构建器
}
```

步骤 2：添加好友状态类，代码如下所示：

```dart
class ContactState extends State<ContactSiderList> {
    // 状态类实现
}
```

步骤3：实例化一个列表滚动控制器 ScrollController，空实现即可。

步骤4：写一个函数判断并显示头部视图，函数名为 _isShowHeaderView。如果不能显示头部则返回一个空容器，函数结构如下所示：

```dart
_isShowHeaderView(index) {
    // 返回消息列表头或者返回 Container
}
```

步骤5：再写一个函数根据定位判断是否显示好友列表头。与上一个函数不同的是这个函数用于判断是否显示，并不返回具体组件，函数结构如下所示：

```dart
bool _shouldShowHeader(int position) {
}
```

步骤6：渲染整个好友列表。这里需要重点提一个 AlwaysScrollableScrollPhysics 类，这个类的对象是要赋值给 ListView 的 physics 属性的，作用是当列表里面的内容不足一屏时，列表也可以滑动。

完整的代码如下所示：

```dart
import 'package:flutter/material.dart';
import 'package:flutter/rendering.dart';
import './contact_vo.dart';

class ContactSiderList extends StatefulWidget {
    // 好友列表项数据
    final List <ContactVO> items;
    // 好友列表头构造器
    final IndexedWidgetBuilder headerBuilder;
    // 好友列表项构造器
    final IndexedWidgetBuilder itemBuilder;
    // 字母构造器
    final IndexedWidgetBuilder sectionBuilder;

    // 构造方法
    ContactSiderList({
      Key key,
      @required this.items,
      this.headerBuilder,
      @required this.itemBuilder,
      @required this.sectionBuilder,
    }) : super(key: key);

    @override
    ContactState createState() => new ContactState();
```

```dart
}

class ContactState extends State<ContactSiderList> {

  // 列表滚动控制器
  final ScrollController _scrollController = new ScrollController();

  bool _onNotification(ScrollNotification notification) {
    return true;
  }

  // 判断并显示头部视图或空容器
  _isShowHeaderView(index) {
    if (index == 0 && widget.headerBuilder != null) {
      return Offstage(
        offstage: false,
        child: widget.headerBuilder(context, index),
      );
    }
    return Container();
  }

  // 根据定位判断是否显示好友列表头
  bool _shouldShowHeader(int position) {
    if (position <= 0) {
      return false;
    }
    if (position != 0 &&
        widget.items[position].seationKey !=
            widget.items[position - 1].seationKey) {
      return false;
    }
    return true;
  }

  // 渲染列表
  @override
  Widget build(BuildContext context) {
    return Scaffold(
      body: Stack(
        children: <Widget>[
          // 列表加载更多
          NotificationListener(
            onNotification: _onNotification,
            child: ListView.builder(
              // 滚动控制器
              controller: _scrollController,
              // 列表里面的内容不足一屏时,列表也可以滑动
              physics: const AlwaysScrollableScrollPhysics(),
              // 列表长度
```

```
            itemCount: widget.items.length,
            // 列表项构造器
            itemBuilder: (BuildContext context, int index) {
              // 列表项容器
              return Container(
                alignment: Alignment.centerLeft,
                child: Column(
                  children: <Widget>[
                    // 显示列表头
                    _isShowHeaderView(index),
                    // 用 Offstage 组件控制是否显示英文字母
                    Offstage(
                      offstage: _shouldShowHeader(index),
                      child: widget.sectionBuilder(context, index),
                    ),
                    // 显示列表项
                    Column(
                      children: <Widget>[
                        widget.itemBuilder(context, index),
                      ],
                    ),
                  ],
                ),
              );
            }),
      ),
    ],
   ),
  );
 }
}
```

16.8.5　Contacts 类

Contacts 类是好友页面的主界面实现类，主要是给 ContactSiderList 类传入好友列表相关的构造器。打开 lib/contacts/contacts.dart 文件，按如下代码编写即可：

```
import 'package:flutter/material.dart';
import 'package:flutter/rendering.dart';
import './contact_sider_list.dart';
import './contact_item.dart';
import './contact_header.dart';
import './contact_vo.dart';

// 好友列表主页面
class Contacts extends StatefulWidget {
  @override
  ContactState createState() => new ContactState();
```

```dart
}
class ContactState extends State<Contacts>{
  @override
  Widget build(BuildContext context) {
    return Scaffold(
      // 主体实现
      body: ContactSiderList(
        // 好友列表数据
        items: contactData,
        // 好友列表头构造器
        headerBuilder: (BuildContext context, int index){
          return Container(
            // 好友列表头
            child: ContactHeader(),
          );
        },
        // 好友列表项构造器
        itemBuilder:  (BuildContext context, int index){
          return Container(
            color: Colors.white,
            alignment: Alignment.centerLeft,
            // 好友列表项
            child: ContactItem(item:contactData[index]),
          );
        },
        // 字母构造器
        sectionBuilder: (BuildContext context, int index){
          return Container(
            height: 32.0,
            padding: const EdgeInsets.only(left:14.0),
            color: Colors.grey[300],
            alignment: Alignment.centerLeft,
            // 显示字母
            child: Text(
              contactData[index].seationKey,
              style: TextStyle(fontSize: 14.0,color: Color(0xff909090)),
            ),
          );
        },
      ),
    );
  }
}
```

16.9 我的页面

我的页面通常包含了用户的个人信息，比如我的相册、我的文件等。同时还有一些软件的管理操作，比如清理缓存、系统设置等。此项目中我的页面效果如图 16-11 所示。我的页面是一个纯 UI 展示的页面，没有数据的添加。布局上采用 ListView 包裹即可，所以实现起来相对简单。

16.9.1 通用列表项实现

由于我的页面相同的项较多，所以这里可以提出一个通用组件用来展示。这个组件命名为 ImItem。

打开 lib/common/im_item.dart 添加 ImItem 组件的实现代码如下所示：

图 16-11　我的页面效果

```
import 'package:flutter/material.dart';
import './touch_callback.dart';

// 通用列表项
class ImItem extends StatelessWidget{
  // 标题
  final String title;
  // 图片路径
  final String imagePath;
  // 图标
  final Icon icon;

  ImItem({Key key,@required this.title,this.imagePath,this.icon}): super(key:key);

  @override
  Widget build(BuildContext context) {
    // TODO: implement build
    return TouchCallBack(
      onPressed: (){
        // 判断点击的项
        switch(title){
          case '好友动态':
            // 路由到好友动态页面
            Navigator.pushNamed(context, '/friends');
            break;
          case '联系客服':
            break;
        }
      },
      // 展示部分
      child: Container(
        height: 50.0,
        child: Row(
          children: <Widget>[
            // 图标或图片
```

```
          Container(
            child: imagePath != null
                ? Image.asset(
              imagePath,
              width: 32.0,
              height: 32.0,
            )
                : SizedBox(
              height: 32.0,
              width: 32.0,
              child: icon,
            ),
            margin: const EdgeInsets.only(left: 22.0,right: 20.0),
          ),
          //标题
          Text(
            title,
            style: TextStyle(fontSize: 16.0,color: Color(0xFF353535)),
          ),
        ],
      ),
    ),
  );
 }
}
```

16.9.2 Personal 类

Personal 类是我的页面的主组件类，由于是纯静态展示，所以继承自 StatelessWidget 即可。打开 lib/personal/personal.dart 文件，按如下代码编写"我的页面"：

```
import 'package:flutter/material.dart';
import '../common/touch_callback.dart';
import '../common/im_item.dart';

//我的页面
class Personal extends StatelessWidget {
  @override
  Widget build(BuildContext context) {
    return Scaffold(
      //列表
      body: ListView(
        children: <Widget>[
          //头像部分实现
          Container(
            margin: const EdgeInsets.only(top: 20.0),
            color: Colors.white,
            height: 80.0,
            child: TouchCallBack(
```

```dart
        child: Row(
          crossAxisAlignment: CrossAxisAlignment.center,
          children: <Widget>[
            // 添加头像
            Container(
              margin: const EdgeInsets.only(left: 12.0, right: 15.0),
              child: Image.asset(
                'images/yixiu.jpeg',
                width: 70.0,
                height: 70.0,
              ),
            ),
            // 用户名及账号显示
            Expanded(
              child: Column(
                mainAxisAlignment: MainAxisAlignment.center,
                crossAxisAlignment: CrossAxisAlignment.start,
                children: <Widget>[
                  Text(
                    '一休',
                    style: TextStyle(
                      fontSize: 18.0,
                      color: Color(0xFF353535),
                    ),
                  ),
                  Text(
                    '账号 yixiu',
                    style: TextStyle(
                      fontSize: 14.0,
                      color: Color(0xFFa9a9a9),
                    ),
                  ),
                ],
              ),
            ),
            // 二维码图片显示
            Container(
              margin: const EdgeInsets.only(left: 12.0, right: 15.0),
              child: Image.asset(
                'images/code.png',
                width: 24.0,
                height: 24.0,
              ),
            ),
          ],
        ),
        onPressed: () {},
      ),
    ),
    // 列表项 使用 ImItem 渲染
```

```
Container(
  margin: const EdgeInsets.only(top: 20.0),
  color: Colors.white,
  child: ImItem(
    title: '好友动态',
    imagePath: 'images/icon_me_friends.png',
  ),
),
Container(
  margin: const EdgeInsets.only(top: 20.0),
  color: Colors.white,
  child: Column(
    children: <Widget>[
      ImItem(
        imagePath: 'images/icon_me_message.png',
        title: '消息管理',
      ),
      Padding(
        padding: const EdgeInsets.only(left: 15.0, right: 15.0),
        child: Divider(
          height: 0.5,
          color: Color(0xFFd9d9d9),
        ),
      ),
      ImItem(
        imagePath: 'images/icon_me_photo.png',
        title: '我的相册',
      ),
      Padding(
        padding: const EdgeInsets.only(left: 15.0, right: 15.0),
        child: Divider(
          height: 0.5,
          color: Color(0xFFd9d9d9),
        ),
      ),
      ImItem(
        imagePath: 'images/icon_me_file.png',
        title: '我的文件',
      ),
      Padding(
        padding: const EdgeInsets.only(left: 15.0, right: 15.0),
        child: Divider(
          height: 0.5,
          color: Color(0xFFd9d9d9),
        ),
      ),
      ImItem(
        imagePath: 'images/icon_me_service.png',
        title: '联系客服',
      ),
```

```
              ],
            ),
          ),
          Container(
            margin: const EdgeInsets.only(top: 20.0),
            color: Colors.white,
            child: ImItem(
              title: '清理缓存',
              imagePath: 'images/icon_me_clear.png',
            ),
          ),
        ],
      ),
    );
  }
}
```